We rejoice and revel in God's word and discovering the breadth and depth of Christ's work for us, yet we continue to discover new and important depths in Scripture. Previous generations misunderstood the Bible about slavery, apartheid and racial discrimination. Dave Bookless shows how we have neglected God's teaching about caring for the environment – which is his creation. The realities of climate change, over-fishing, global pollution, ozone depletion, and soil instability through over-grazing and deforestation are increasingly apparent, yet we fail to recognize that they arise because we have ignored our God-given responsibilities to his world. Paul is explicit that Christ's death on the cross reconciled all things to the Father, not merely human beings. God has given us a Book of Works (creation) as well as a Book of Words (the Bible). We will be misshapen Christians unless we read both – Christian couch potatoes bulging in the wrong places. *Planetwise* takes us helpfully, clearly and powerfully through the whole Bible from Genesis to Revelation, pointing out the amount and relevance of what it says about why and how to look after creation – as God commands.
Professor R. J. Berry, former President of the British Ecological Society

In an accessible and easy style, Dave Bookless introduces us to a biblical and practical response to climate change and environmental stewardship. This is a must-read for all disciples of Jesus committed to God's mission in the world.
Revd Canon Tim Dakin, General Secretary, CMS

In *Planetwise*, Dave Bookless resists the temptation of tackling only the topical issue of climate change. Instead, he gives a clear, Bible-based overview of our place in God's world, which challenges and changes our thinking at the deepest level. I've had the privilege of seeing A Rocha's work at first hand over the last few years, and Dave's track record in practising what he preaches means this book has real integrity. It's accessible, biblical, practical and passionate. Do read it and pass it on to your friends!
Dr Ram Gidoomal CBE

Dave Bookless is passionate about his faith and about the environment. In this book he offers us a chance to explore a massive global issue of our generation, in both theological and practical ways.
Revd Nicky Gumbel, Vicar of Holy Trinity Brompton and pioneer of the Alpha course

In this compelling book, Dave Bookless challenges us to think radically and biblically about one of the biggest challenges that faces us all today – the future of the planet. While it has become fashionable to concentrate our minds on climate change, I believe Dave is right to argue that the problem runs much deeper – the whole way we live our lives today. He convincingly argues that our entire lifestyles need ch

All Christians who really want to explore the entire meaning of the Bible should read this book. For too long we have been taught only part of the Bible, missing the need to move beyond what the Bible tells us about our individual relationship with God. The whole gospel is about Jesus transforming not only our relationship with God, but with other people and the world around us. Our task as human beings is to bring the good news of Jesus to all of creation in both word and deed. If we don't live our lives in a way that worships the God of the whole Bible, how can we expect people to hear the good news?

I have personally come to the same view as Dave over the last few years, and have been wanting to write such a book as this myself. Now Dave has done the job for us. He brings to this book accessible theology as well as practical lifestyle ideas. He explores his subject with humility and a deep recognition that each individual's situation is very different. He has encapsulated all the necessary thinking about the sustainability of our Western lifestyles, and hopefully will challenge you as much as he has me about how and why we live as we do.

After reading this book, I hope you will feel as liberated as I do about a new way of living – one in which we get our personal relationship with God right as well as our relationship with our fellow human beings and the planet. I hope you will feel challenged to alter your lifestyle, and I am convinced you will want to do so after prayerfully considering this book's message to all of us.
Andy Reed MP

As President of Tearfund, I have seen how a changing environment harms the world's poor. Sometimes people argue that we should care for the poor rather than the planet. In his book, Dave Bookless shows how this is a false and unbiblical distinction. For the Bible reveals a God who cares for all creation – human and non-human – and calls us to do the same. We cannot help the poor without being *Planetwise*. Do read this book, buy it for friends, and be prepared to be challenged.
Dr Elaine Storkey, theologian and social scientist

Dave's passion for God's world shines through on every page of this well-written and inspiring book. As we look at how humans have treated this world that God has entrusted into our care, we desperately need to discover afresh how the Bible speaks into our current situation. Dave's in-depth biblical knowledge and honest reflections on how they live as a family will both challenge and equip us to play our part in bringing hope to this world and all its inhabitants.
Ruth Valerio, author of L is for Lifestyle: Christian Living that Doesn't Cost the Earth

DAVE BOOKLESS

PLANET WISE

DARE TO CARE FOR GOD'S WORLD

ivp

INTER-VARSITY PRESS
Norton Street, Nottingham NG7 3HR, England
Email: ivp@ivpbooks.com
Website: www.ivpbooks.com

Illustrations by Tiphanie Goldspink

First published 2008

British Library Cataloguing in Publication Data
A catalogue record for this book is available from the British Library.

ISBN: 978-1-84474-251-6

Set in Dante 11/13pt
Typeset in Great Britain by CRB Associates, Reepham, Norfolk
Printed in Great Britain by Ashford Colour Press Ltd, Gosport, Hampshire

Inter-Varsity Press publishes Christian books that are true to the Bible and that
communicate the gospel, develop discipleship and strengthen the church for its
mission in the world.

Inter-Varsity Press is closely linked with the Universities and Colleges Christian
Fellowship, a student movement connecting Christian Unions in universities and
colleges throughout Great Britain, and a member movement of the International
Fellowship of Evangelical Students. Website: *www.uccf.org.uk.*

Contents

Foreword: Making a difference!

Dave Bookless was concerned that there was so little green space in his parish. He noted that the Southall Regeneration Partnership Report of 1998 concluded that there was 'a lack of greenery, open space, clean air and environmental awareness – all of which contribute to a lack of confidence and pride in the area'.

He was disturbed by the way that a large plot of land known locally as the 'Minet site' had been so neglected and abused. It had become an illegal fly-tipping site, the home of an enormous unregulated car-boot sale that resulted in tons of litter, and a late-night track for local motorbike racing: a muddy quagmire of refuse and waste.

Dave and his team became key players in the formulation of plans to turn most of the site into a country park that would cater for the needs of both local wildlife and people. After many battles and tough times, they eventually gained planning permission for the creation of Minet Country Park on 30 May 2002. Gradually they gathered a huge group of volunteers, received grants, and gained solid support for the project from the whole community.

June 2002 saw the beginning of the transformation. Dave partnered with the Christian ecology movement A Rocha, and together they produced an Ecological Impact Assessment on the proposed works, acting as Ecological Advisors throughout.

They oversaw the habitat creation and excavation of wetland scrapes, the fencing of areas for the protection of ground-nesting birds, and the removal of twenty lorry-loads of illegally dumped rubbish. They developed a water trail, ringed 600 birds of twenty-three species, and cared for the twenty-two species of butterfly, including the Small Copper. They built strong links with four local primary schools, and developed a curriculum-linked programme for environmental education. They started after-school environment clubs and holiday play-schemes on the site for sixty local children, and organized family picnics, insect safaris and wildlife walks. The team conducted twenty-three assemblies in eleven local schools and created resources for a purpose-built 'floating classroom' on the Grand Union Canal beside the site.

It was my great privilege to take part in the official opening of Minet Country Park. It was a parable of redemption, a great witness for Christ, and a sign that Christian mission is credible when it takes our theology of a Creator God seriously. As I meandered around the site that hot summer afternoon, I heard the birds singing and saw skylarks, kingfishers and woodpeckers flying free. Butterflies fluttered through the long grass. A conservationist called me over to wonder with him at a wild orchid growing by a pond. I looked at the distant line of planes queuing to land at nearby Heathrow, and marvelled that even here, near the disused gasworks, this urban wasteland had become another Eden.

The Living Waterways project is only one small aspect of Dave's extensive ministry with A Rocha UK, and he already acts as consultant to many other pieces of 'eco-mission'. Dave is a contemporary evangelist, taking the good news of Jesus into situations many vicars would never get near, because he carries with him the urgency of our global crisis.

That's why this book is such an important contribution to the Christian debate about the environment. Dave has not written this from some ivory tower of academic learning. No, the theology that underpins this book has been forged through a life and ministry totally committed to developing a Christian response to the environmental crisis.

Dave's theology has led him towards a fresh and innovative style of contemporary Christian ministry which draws many committed greens to work alongside him. He is doing much to reclaim the ground which we have lost to the followers of 'eco-spirituality', and his life is a witness to living out a credible Christian eco-spirituality personally. This book comes out of a ministry developed in the midst of tough places, a multicultural parish, a communal commitment to live simply, and a personal hunger to discern God's perfect will.

Some churches proudly claim that they recycle their bottles or abandon their cars to go to church one Sunday per year. They are to be commended, but this kind of tokenism falls far short of a credible response to the cataclysmic effects of global warming. Dave's ministry models something far more relevant and profound. This book teaches that if we are to embrace a kind of Christian eco-spirituality which has a vision to save the created order, it will demand real sacrifice and a very different way of life.

Our stewardship of the planet will one day be judged. And that means a change of perspective – a new look at how we live, and a new commitment to working with the Lord in renewing the planet.

Our understanding of creation must begin with a sense that it is God's and not ours. He didn't leave it and walk away; his presence infuses it. The choice about how you live is yours. But ultimately, the judgment is his.

The contemporary church urgently needs Dave's insights if it is to refocus its worship and prayer towards a new awareness of the Creator God. Dave's theology demonstrates how genuine ecological action must flow out of the core of our believing rather than being some kind of add-on optional activity to ease our consciences.

The world produces two million tonnes of rubbish every day. Half a billion tonnes of oil are spilled every year through accidents, dumping and leakage. Six and a half million tonnes of refuse, including toxic and non-biodegradable waste, are discharged every year into the world's oceans. So what basic principles should be guiding us in how we live out our faith in this increasingly polluted planet, and how should Christians respond to this environmental crisis?

This book is the starting-point. Read on.

Revd Dr Rob Frost
November 2007

Acknowledgments

Writing this book has been a bit like exploring an iceberg. Ninety per cent is hidden from view, and that includes the enormous debt I owe to all who have opened my eyes to God's word and God's world over the years. Here I can only name a few of those without whom this book would have been impossible: Peter and Miranda Harris who inspired me to begin a journey with A Rocha that has led to where I am now; Sian Hawkins, Steve Hughes, Pete Hawkins and other colleagues in A Rocha UK for giving me the time to write and for many insights and encouragements; Eleanor Trotter at IVP for keeping me to task; Tiph Goldspink for her illustrations done whilst juggling three children and other work; Anthony and Pauline Hereward for opening their home for me to write in; Rob Frost for writing such a kind foreword whilst he struggled with cancer during the final weeks of his life; Alice Amies, David Chandler, Sarah Walker and all those others who read various drafts and weren't afraid to put me right. Finally, I want to thank Anne, my soul-mate, fellow-conspirator, constant encourager and best critic, and our daughters Hannah, Rebekah, Rosie and Naomi-Ruth for keeping my feet on the ground. This book is dedicated to my mother Rosemary and my father Guy (who died in 2006 but whose life of humble Christian witness continues to inspire me), and to the glory of God – Creator, Sustainer, Redeemer.

Dave Bookless
Southall, December 2007

Introduction:
Planet Earth – why bother?

Many people say climate change is the biggest threat our world faces today. I beg to differ. It's not that I doubt the scientific consensus on the threats posed by melting ice-caps, changing weather systems and warming oceans. Nor am I blind to the terrible effects these will have, indeed are already having, on wildlife, the poor, and ultimately all of us.

The reason is this: climate change is a symptom of a far bigger problem. Imagine if you switched on your television and heard that science had discovered a 'cure' for climate change: a magical solution to absorb all the excess greenhouse gases. Imagine that the atmospheric clock was turned back so that 200 years of industrial pollution were no longer going to cause ice-caps to melt, oceans to expand, forests and coral reefs to die and hundreds of millions of people to be forced to migrate. Would we then have a perfect world with no environmental problems?

Sadly, the answer is 'no'. We would still be facing enormous environmental problems. Forests would still be destroyed, oceans over-fished, resources over-exploited. Our countryside would still be disappearing under mountains of waste. Dangerous pesticides and chemicals would still be causing huge problems to ecosystems and human health. People in rich nations would still be consuming enormous amounts of the earth's resources and living energy-hungry lifestyles, while those in poor countries would struggle just as hard to find food and water. Many species of wildlife would still be driven to extinction as human populations sprawled into their habitats.

Climate change is simply the most obvious symptom of a much, much deeper sickness. At the heart of it is this: as human beings we have got our relationship with the planet all wrong. It is not just that populations are growing and energy-hungry lifestyles increasing, but that we have been living in a way that simply cannot continue. We cannot solve this problem simply by better technology and a few hard political choices. It goes deeper than that, right to the heart of who we are. We need to rethink not just how we treat the planet and its creatures, but who on earth we think we are as human beings. That, in essence, is what this book is about.

Why should we care?

There are many reasons why people get involved in environmental issues. For some it may be a love of wildlife, gardening or beautiful countryside. For others it is the latest big cause to follow, a way of making the world a better place. For an increasing number it's quite simply fear: if we don't do something now, we're all going to suffer in years to come. For many, including Christians, caring for the environment is an issue of justice. It's about the millions of the world's poorest people who are already suffering the results of a changing climate.

Christians, however, have to start with the Bible. Imagine for a moment that no human being was going to suffer as a result of environmental mismanagement. Would we still care about the planet – about disappearing wildlife, polluted skies and poisoned seas? Is the Christian gospel simply about rescuing people from a dying world and telling them the good news of heaven, or does God care about how we live on earth now? Do other species, and the earth itself, matter to God? Or are they simply here for human beings to enjoy?

These questions must be our starting point. There is a real danger that we get caught up in all the hype of the latest good cause without stopping to look at how this fits into a biblical view of the world. This book aims to take a step back and let the Bible tell its own story in terms of how God sees this world and our place within it.

Over the past ten years I have talked with several hundred different church groups around the UK about the Bible and the environment, everywhere from Aberdeen to Penzance, from St David's to Walton-on-the-Naze. I've also been privileged to speak at conferences in four continents on the same subject. What has become clear is the great variety of views about how the environment relates to Christian faith. At the risk of stereotyping, I've grouped these views under four broad themes. As we look at these, please try and see where you fit in.

Insidious: *Ecology and environmental issues are a bit dodgy, and Christians should keep well clear.*

Some are concerned that the New Age movement has infiltrated and taken over the green movement. Images come to mind of tree-hugging pagans dancing at full moons, worshipping at ancient sites and indulging in fertility festivals. In fact, the New Age movement is a varied mix of groups and ideologies, some drawing their ideas from eastern Hindu and Buddhist concepts, some from modern versions of ancient pagan ideas, and some from occult sources.

You will certainly find environmentalists who talk of the earth as 'Gaia', the ancient mother goddess, and who practise nature worship. You will also find some who are anti-Christian, believing Christianity justifies exploiting nature and is therefore to blame for the environmental crisis.

However, this is a strange reason for Christians to avoid environmentalism. The environment is created by God, not the New Age movement. It's almost like saying Christians shouldn't listen to music because some musicians have dubious beliefs. That may be true but it misses the whole point! Music, like the environment, is God's good creation. Actually, many environmentalists are not into New Age, pagan or occult ideas. Many are agnostic or atheist, and a growing number are committed Christians.

Imagine if Jesus refused to mix with people he disagreed with: dishonest tax-collectors, outcast prostitutes, self-righteous Pharisees, argumentative fishermen. He would have had very few disciples. The environmental movement certainly includes people with a very different view of the world from Christianity, but also many who are openly searching for spiritual reality. It's a compelling reason for Christians to get involved.

Irrelevant: *Caring for the earth is not important for Christians. The gospel is about saving souls, not saving seals.*

This is a view that is strongly held in some churches, and I often hear it expressed through questions when I visit:

- 'Isn't the gospel about spiritual, not material matters?'
- 'Doesn't God care about our souls, not our bodies?'
- 'Shouldn't we just be focusing on evangelism rather than worrying about the planet?'
- 'Shouldn't our minds be on heaven and not on earth?'
- 'Isn't God going to destroy this earth anyway?'

I will address all of these questions later. Here, however, I simply want to question the mindset behind them, because there's a danger that they are ultimately based on an unbiblical understanding of reality.

At the time the New Testament was written, there was a battle of ideas between the dominant pagan Greek philosophy and the new Christian ideas founded in Old Testament Jewish thought. At its heart, the battle was over whether ultimate reality is purely spiritual. Are human beings divine souls trapped in physical bodies, or are our bodies part of who we really are? The Bible is very clear. We are not merely spirits or souls, and our material bodies are vitally important. In 1 Corinthians 15, Paul reminds us that Jesus rose from death with a physical body, that after death we too will have physical bodies, and that Christianity falls apart without this. The biblical view is that mind-body-spirit together make up who we are. You will not find a single New Testament passage that speaks about 'saving souls', because Jesus was not interested in disembodied souls! He was interested in whole people; he healed physical and mental illnesses as well as forgiving sins. He taught us to pray for God's kingdom 'on earth', not just in heaven.

There's an old song that goes: 'This world is not my home, I'm just a-passing through.' It springs from the experience of Christian slaves in the southern United States and represents an understandable reaction to the terrible suffering and injustice of slavery. However, it's not what the Bible teaches. As we will see, this earth *is* our God-given home, and the Creator cares about his house-guests' behaviour.

Finally, the question about the earth being destroyed is an important one (which is why the whole of chapter five is devoted to it). For now, it's worth mentioning that this isn't what most Christians have believed down through the ages. It's a relatively modern view that grew up alongside the Industrial Revolution. Perhaps the link is unfair, but I sometimes think it's been very convenient to believe the earth is disposable, at a time when we've been exploiting and destroying its resources as never before.

Incidental: *I'm glad somebody's caring for the planet, just as long as it doesn't have to be me!*

In my experience, this is the majority view of most Christians, most of the time. If I'm honest, I've been there too. Life is full of important issues and we can't get involved in all of them. It's great that some Christians are called to get involved in caring for the environment. Good for them, but it isn't for everybody!

> *This earth is our God-given home, and the Creator cares about his house-guests' behaviour.*

After all, there are many other special interest groups. Not everybody is called to be part of Christians in Sport or Christian Surfers. Surely caring for the environment is just for those who are into gardening or bird-watching, or those who go weak at the knees at the sight of dolphins and baby seals?

Actually, it's not like that. Of course there are many areas of life that only a few people are called to get involved in. However, there are also areas right at the heart of the Christian faith which anybody who is a follower of Jesus must take on board. Take prayer, for example. Can you imagine a Christian saying, 'No, I don't bother with prayer. It may be for some people, but it's not for me. You go ahead and pray, but please don't talk to me about praying'? Of course not! Whether we're new Christians or have been missionaries for fifty years, whether we think we're 'good' at prayer or are just beginners, whether we actually pray regularly or not, we all know that prayer is an essential part of the Christian life. Some people may be called to a special ministry of prayer – intercessors or prayer warriors – but everybody is called to pray.

As I've re-read the Bible, I have come to realize that caring for God's creation is not insidious, neither is it irrelevant or merely incidental. Instead, caring for the earth and its creatures is a core part of what all Christians are called to.

Integral: *Concern for the whole of God's creation is fundamental to the God of the Bible and to his purposes for human beings.*

Just as all Christians are called to pray, meet together, study God's word, and share the good news, so caring for creation is essential to following Jesus Christ. It's not an optional extra, but part of the core of our faith. I don't mean that everybody is called to live in a tepee, campaign on behalf of whales, or go vegan! Rather, following Jesus means looking at God's world with new eyes.

Every now and then there are major shifts in Christian thinking, as we wake up to biblical truths which our culture has prevented us from seeing. Two hundred years ago Christians like William Wilberforce changed the way people thought about the slave trade. Until then, slavery was permitted, and even justified from the Bible. However, Wilberforce, Shaftesbury and others allowed the Bible to challenge the prejudices of their culture. They recognized that when Paul wrote, 'There is neither Jew nor Greek, slave nor free, male nor female, for you are all one in Christ Jesus' (Galatians 3:28), it meant a new way of looking at people. The new community of Christ-followers were equal in God's sight and slavery could no longer be condoned.

I believe we are at one of those moments today. It's as if we're removing a pair of tinted glasses that have coloured our whole view of life. Our culture, especially our Western, urban, industrial, consumer culture, has surrounded us so effectively that we've failed to notice the plain message of the Bible on creation and our place within it. Now, at last, we're being forced to think again as we face up to the damage our way of life has been causing this planet.

Many of us only come to the Bible asking questions like, 'What does this tell *me* about *my* relationship with God?' We tend to see the Bible as being all about people. Actually it is all about God. Alongside the familiar material about God's dealings with humanity, there's a huge amount about God's dealings with the earth which we've tended to overlook. Most of us have failed to ask what the Bible says about the planet, about God's relationship and ours with it.

This is a major shift to make. It's similar to when the famous astronomer, Copernicus, first realized the earth revolved around the sun. Many people found this threatening and regarded Copernicus as a heretic, because they misread the Bible as saying that the earth was at the centre of all things. Today our perspective needs to shift just as radically. We need a change of worldview. We are not the only focus of God's creative and saving love. Rather, God cares about all that he has made. We urgently need to recognize that the earth and the creatures with which we share it are not merely the stage on which we act out our relationship with God. They are characters in the story themselves.

This book does not attempt to pick out a few Bible verses to support these views. Rather it tries to look at the big picture, the whole story of the Bible, and ask some basic questions. The next five chapters do this using a helpful framework suggested by Dr N. T. (Tom) Wright, bishop of Durham. He writes, 'The early church saw history as a five-act play, with creation, fall and the story of Israel as the first three acts, and the drama reaching its climax in the fourth act, the events concerning Jesus of Nazareth. The early church itself was living in the fifth act, where the actors are charged with the task and responsibility of improvising the final scenes of the play on the basis of all that has gone before.'[1] This is a wonderfully simple summary of the great drama of God's dealings with creation, played out through history and retold in Scripture. The five acts, paralleled by the next five chapters of this book are, once again:

1. Creation
2. Fall
3. Israel
4. Jesus
5. The present and future age

While the five acts brilliantly sum up the story of the Bible, we will find that they frequently overlap. For example, the New Testament explains and expands on the Old Testament's understanding of creation and the fall. In a similar way, our understanding of the final act, the future of the earth, is not only found in the last few books of the Bible. Thus this book does not start with Genesis and work through to Revelation. Rather, the five acts are like five great themes in a complex orchestral piece, expanding on each other, overlapping and reprising. Like a great musical work, the total is far greater than the individual themes and only makes sense when they are all taken together.

The natural world is by no means a minor player in this great musical drama. As we will see, creation is significant in every single act of the story. This is the story of God and the whole of creation.

Questions

1. Do you agree that 'as human beings we have got our relationship with the planet all wrong'? What evidence can you find to support this view?
2. Where would you place yourself in the spectrum of Christian attitudes to environmental issues: are they insidious, irrelevant, incidental or integral? Why?
3. The chapter concludes that the Bible is not just the story of God and humanity, but 'the story of God and the whole of creation'. Why do you think Christians have often only emphasized the 'human' elements of the Bible, and neglected the wider creation?

1 Creation calls

The drums roll, the curtain slowly parts and the lights gradually fade, leaving an expectant darkness. On the stage itself all is silent, dry ice swirling around formlessly. Suddenly a voice off-stage declares: 'Let there be light!' and the stage is lit by an eerie glow.

Of course, we can only imagine exactly how the creation of all things took place. The account we have in Genesis 1 and 2 is inspired by God but there were no human witnesses until the end of the first scene. The Bible's creation account is deliberately high drama, full of powerful and beautiful language and imagery. It's a story many of us know very well, but one where we often focus on certain aspects and miss the big picture.

We are not going to spend time on whether Genesis 1 and 2 give an exact literal description of how God made the world. The debate over whether this is accurate history or designed to answer profound questions through symbolic language will not be solved here. Arguments over evolution and creationism have divided Christians for many years. In fact, this whole argument has become a huge distraction. The most important questions Genesis answers are not about 'How did we get here?' but 'Why are we here?', not 'How did God create the world?' but 'Why did God create this world?'

By starting with 'why?' rather than 'how?' we will find that the biblical text helps us to understand the whole purpose of creation, as we ask:

- What can we learn about God from creation?
- What can we learn about creation itself, in particular about its relationship with God and humanity?
- What can we learn about being human and our place within creation?

God: beyond creation

If you compare the Bible's creation account with other ancient creation stories, one difference stands out above all others: God creates out of nothing. In other accounts, either raw material already exists, from which creation is shaped, or else the universe emerges or emanates from a Creator. Not so in Genesis 1. Before creation existed there was no cosmic soup, no light, no alien life – there was only God.

God is therefore of a different order of reality from the created universe, and it is difficult to describe him using human language. Yes, God is personal, but so much more than a human being. Yes, God is powerful, yet not even the power within the sun can touch his power. Creation emphasizes God's uniqueness and otherness.

This is important at a time when many worship God 'in nature'. Creation cannot tell us all there is to know about God. However beautiful, mysterious and inspiring the universe may be, we must never think that it is God. However beautiful a dolphin may be, it is not part of God. However amazing planet Earth may be, it is not God. Talk of 'Mother Earth' or 'Mother Nature' confuses the created and the Creator. If the earth is amazingly well maintained, it is because God is maintaining it, not because the earth is God. Ultimately, to worship the creation or any part of it breaks the first of the Ten Commandments: 'You shall have no other gods before me' (Exodus 20:3). It is also idolatry: worshipping a created object instead of worshipping God.

God: revealed through creation

While we must be careful never to confuse God and creation, there is also another side to the story.

Jonathan, aged six, comes rushing home from school proudly clutching a painting. He spreads it out on the kitchen table for mum to look at. It's a self-portrait, although the hands have too many fingers, the legs are stick-like lines, and the head is a huge circle with an enormous red smile. However, the picture tells us a lot about the artist. This is a young child whose skills are still developing. He loves bright bold colours and feels happy about himself. No doubt a child psychologist could tell us more, and a forensic detective would find smudged fingerprints and positively identify the artist!

God is the great artist, and the world his canvas. The world communicates important messages about God to us. Like Jonathan's painting, we can look at God's world and find clues to the artist's personality and character. In addition, God has quite deliberately left clues about who he is: creation is his first chosen means of telling us about himself.

In Romans 1:20, we read, 'For since the creation of the world God's invisible qualities – his eternal power and divine nature – have been clearly seen, being understood from what has been made, so that men are without excuse.' God's power and character are displayed throughout nature. His eternal power is clear to all: controlling the forces at the heart of the universe, the energy in the heart of each star. Yet equally there is God's attention to detail in the delicate tracery of a spider's web, the infinitely different patterns of each snowflake, or the colours of a forest in autumn. God is a God of beauty and order as well as power. Why not stop reading at this point, go outside or look out of a window, and find something that speaks to you of God's eternal power and divine nature?

Today, many people are seeking spiritual reality but see Christianity as irrelevant to the questions they are asking. Creation is a place where they encounter something of that spiritual reality, sensing a greater presence, experiencing the ordering of seasons and tides, or feeling a deep empathy with other creatures. Sadly, Christians have often dismissed such experiences as nature worship. Yet surely this is exactly what Romans 1:20 means by seeing God's invisible qualities through what has been made? Creation is a natural place of encounter with God. What a tragedy that those struggling to make sense of God's fingerprints in creation often find Christians closing the door in their faces.

Creation can also help those who are already Christians understand God better. For instance, creation shows us how relational God is. As scientists study the world, they discover its incredible interdependence. The planet's systems are amazingly fine-tuned, from the way gases in the atmosphere balance each other, creating the perfect conditions for life, through to the millions of microscopic creatures that give fertility to the living soil. The science of ecology is all about how vitally important these links are. Plants, animals, human society and everything else relate to, and are dependent on, one another.

This should not surprise us! The universe is interdependent and relational because it was made by a relational God. All creation (ourselves included) flows from the love that existed before time between the three persons of the Trinity: Father, Son and Holy Spirit, who were all involved as co-creators. In John 1, we read that before the world was created, Jesus, the Word, already existed and it was through his word of command that God created everything. In Genesis 1:2, we read of the Spirit of God hovering over the waters, and in Genesis 2:7 God breathes his breath or Spirit into the first human being. In the third century, Irenaeus wrote that God crafted the world with the two hands of his Son and his Spirit.[1]

As God creates, so God relates. He knows each corner of creation intimately and cares for it. There is a wonderful Bible passage where God shows Job the wonders and mysteries of creation, pointing out that he cares for and sends rain even 'to water a land where no man lives, a desert with no-one in it' (Job 38:26). It is not only humans, but also the parts of creation that have nothing to do with humanity, which are important to this relational God. Environmentally, this matters because it demonstrates that creation is not only about people, and God's relationship with creation is independent of our relationship with it.

Knowing that God has made a relational universe also reminds us that human beings are not designed to live in isolation. We've been created to relate to God, to each other and to the natural world. We depend on plants, animals and the systems God has created to provide oxygen to breathe, water to drink and food to eat. Without these, we will die. We are also interdependent, with the rest of creation, upon God. Nothing could be more dangerous than the illusion our modern societies encourage that we can get by on our own. God has made us to depend on himself, each other and the whole of creation.

God: committed to creation

Sometimes people speak as if God's involvement with the universe ceased after creation. Some call this 'the Divine Watchmaker': a God who wound it up, set it going and then moved on, leaving it to unwind gradually. However, God's relationship with this created world is far closer than this. It's nearer to that of parents with a young child they love deeply. Attached by an unbreakable cord of love, they cannot abandon their loved one. How then could a relational God, who created in love, walk away and abandon his creation?

The biblical story reveals that God's relationship with creation did not stop after the six days of Genesis 1. God is its sustainer, continuing to uphold, care for and renew his creation. The Psalms talk about this over and over again. 'O LORD, you preserve both man and beast,' says Psalm 36:6. Psalm 65 lists God's care for the land in terms of rain, growth and harvest. Psalm 74:16–17 says the days and seasons are at God's command.[2] If the Psalms were the songbook of God's people, it's clear that what inspired their singing more than anything else was God's creation. God loves this world deeply and is involved with it intimately. It is a two-way relationship that includes animate and inanimate objects: mountains, rivers, lightning and rocks as much as animals and people.

As in creation, Father, Son and Spirit are each involved in sustaining God's world. According to Colossians 1:17, 'all things hold together' in Christ. God's Spirit also fills and 'inspires' the universe: 'Where can I go from your Spirit? Where can I flee from your presence?' asks the psalmist. 'If I go up to the heavens, you are there; if I make my bed in the depths, you are there. If I rise on the wings of the dawn, if I settle on the far side of the sea, even there your hand will guide me, your right hand will hold me fast' (Psalm 139:7–10). The Bible is clear that God is deeply involved in the seasons and cycles of nature, in processes from the sub-molecular to the pan-galactic. He is involved because he made the universe in love and continues to uphold it in his love and power.

We live in a time of increasing fear about the earth's future. Respected scientists and economists tell us we are over-using the earth's resources and unleashing forces beyond our control. Some predict a complete collapse of human civilization, with billions of people dying through drought, famine, flood and disease. Others believe we have a chance if we make rapid and major changes to the way we live. It is all too easy to lose all hope for the future.

Our small actions might seem too little, too late, on their own, yet God is committed to including us in his plans. He can take our small efforts and weave them into his purposes in sustaining and renewing the earth.

Yet as Christians we can have hope. 'Sustainability' is the new holy grail. It's about living in a way that leaves enough of the earth's resources for future genera-tions and other species. We believe God is committed to looking after this earth. In the search for sustain-ability, he is the Sustainer. That doesn't mean we sit back and do nothing. As we'll see later, God has given human beings a special responsibility in caring for the planet. However, it means we don't need to despair. Our small actions might seem too little, too late, on their own, yet God is committed to including us in his plans. He can take our small efforts and weave them into his purposes in sustaining and renewing the earth.

Creation: good!

As a small child, I remember adults warning me about the 'big, bad world'. They were simply trying to prevent me being from over-adventurous, but the image it left was of a dangerous, hostile world full of lurking evil. As I grew older, this pulled me in two directions. On the one hand, I saw plenty that was wrong with the world – pain, suffering and injustice – and I believed that only in God's kingdom (which I saw as the church) could things be put right. On the other hand, I loved wild nature: the pull of a majestic mountain crying out, 'Climb me!', the sound of waves breaking on a rocky shore, the sight of wild geese migrating in formation against a setting sun. I struggled to reconcile these two pictures: of the world as evil and the world as beautiful and good.

When we read the Genesis creation account, the first thing that strikes us is that creation is good. After making the land and seas, God stopped, looked at what he had made, and 'saw that it was good'. He said the same after making plants and trees, creating the sun and the moon, and fish, birds and animals. Finally, after making people, he stopped and looked at everything he had made, 'and it was very good' (Genesis 1:31). It is astonishing how often we overlook the simple truth that creation is good. It reflects the goodness and character of God. He made it, he loves it, and that settles it: we should love it too. Physical matter matters, because it is important to God. Whatever has happened since, God made a good world.

We must, therefore, reject the kind of separation I grew up with between God's kingdom (good) and the world (bad). There is no such separation. Everything in all creation was made good. Even those things we struggle to see as good had an original good purpose in God's plans. In fact, it would be impossible for a good God to create anything that isn't good.

Nor can we say that creation has lost all its goodness since sin's entry into the world. Just read the Psalms! They celebrate the goodness of creation long after evil entered the world. In the New Testament, Jesus saw creation as a treasure-house to illustrate God's goodness and character. As Christians, we're told clearly that 'everything God created is good, and nothing is to be rejected if it is received with thanksgiving' (1 Timothy 4:4).

Let us rediscover that the gospel, the *good* news, does not begin with Jesus' birth. It begins with the good earth that God made through Jesus. Let us celebrate again that creation in all its richness is the wonderful gift of a good God.

Creation: speaking of God

I have many friends who love wildlife, including Bill and Peter, who join me in scientific bird-ringing studies by catching wild birds and attaching a tiny aluminium ring with a unique reference number to their legs. The information gained helps us understand bird migration, distribution and population changes, and protect wildlife better. In the UK, bird-ringing is controlled by the government under licence to the British Trust for Ornithology and great care is taken to ensure the well-being of the birds. They are mainly caught in almost invisible mist-nets, into which they fly and in which they get tangled. After being carefully extracted, they are identified by species, age and sex; wing length, muscle and fat health are measured, and they are weighed. Finally the birds are released to carry on feeding, nest-building or migrating.

Most of my bird-ringing friends would not call themselves Christians, yet time and time again we have had conversations about the sheer wonder of these tiny feathered creatures. Birds are amongst the most colourful and musical members of the animal kingdom. From a purely scientific perspective, colour and song perform useful roles: attracting mates, competing with rivals, establishing and defending territory, and hiding from, or scaring off, predators. Yet the colour and song of birds go beyond this. Mike Brooke, curator of ornithology at Cambridge University's Museum of Zoology, admits, 'We lack any overarching theory for why birds are coloured the way they are.'[3] There is an almost universal emotional connection that people sense when they see the bright colours of a kingfisher or the iridescence of a peacock's feathers, or when they hear a skylark or a nightingale singing. The colour and song of birds speak to us of nature's awesome beauty and of the character of the creator God.

The great joy of bird-ringing lies not only contributing to valuable scientific research, but also in handling tiny delicate creatures that have such amazing lives. In the summer, we catch small songbirds, such as garden warblers and whitethroats, weighing between 10 g and 20 g, which migrate from sub-Saharan Africa to breed in Britain. The methods they use to navigate are still being discovered but many seem to use the stars. Sometimes we catch a warbler with a ring from a previous year. It may well have found its way from west London, down through Europe, across the Mediterranean Sea, either across the Sahara or around the west coast of Africa, to its eventual destination in Senegal or Gambia. After several months enjoying sunshine and abundant food, it has made the same journey in reverse, avoiding rainstorms and hunters' guns, and returning exactly to its starting place.

Even a hardened atheist struggles to hold back a sense of awe at this. Bird migration is little short of a miracle, and like so much else in creation, it points to God. These tiny birds are amongst the most eloquent evangelists I have ever met. Without words, they proclaim the glory of God.

Looking at birds gives a tiny insight into the amazing variety God has made: from tiny bee hummingbirds smaller than a thumb, with hearts beating at a thousand times a minute, to flightless emperor penguins raising their chicks at 50° below zero; from peregrine falcons diving at breathtaking speed, to enormous ostriches with their football-size eggs. In terms of colour, shape, size, skills and migration strategies, they tell us about an amazingly creative and imaginative God . . . and that's just the birds! There are perhaps 10,000 species of bird out of perhaps 1.8 million living species identified so far (more than 300,000 of them beetles!). According to many estimates, there may be around 10 million species altogether, most still undiscovered. God is creative beyond our wildest imagination, and all biodiversity reveals something of his existence and nature.

Creation speaks fluently and eloquently about God. The earth and the life forms it contains teach us that God exists and that he communicates with us through what he has made. In Psalm 19, we read, 'The heavens declare the glory of God; the skies proclaim the work of his hands. Day after day they pour forth speech; night after night they display knowledge. There is no speech or language where their voice is not heard' (verses 1–3). The psalmist makes it clear that creation speaks of God in a way that goes beyond language and right to the heart. God speaks through two books: his word (the Bible) and his works (creation). We need both together to understand what God is really like. Without the Bible, we may see God's fingerprints in creation but we cannot form a full portrait: they are like fragments in a kaleidoscope. Yet without creation, the picture of God we get

from the Bible is also incomplete. Creation illustrates and illuminates
the truths about God that we read in the Bible.

There are many occasions where God uses creation to speak to his
people. When Jonah tries to run away from God, a large fish
swallows him before later disgorging him onto dry land. Later, when
he is sulking, God speaks to him through a vine that grows and
spreads to give him shade, and then through a parasite that eats the
vine so it withers up. Elsewhere, when people won't listen to God's
voice, he even speaks through a donkey![4]

One of the clearest examples is in the story of Job. When all that he
has in life is stripped away and Job is left friendless and penniless, God
does not speak to him through human wisdom, religious observance,
or Scripture. He speaks through creation, asking him to consider the
mysteries of the galaxies and of weather systems. He takes him on a
tour of his creation, studying mountain goats, wild donkeys, oxen,
ostriches, horses and hawks, before finally introducing the monstrous
behemoth, the hippopotamus, and the fearsome leviathan, the
crocodile. Somehow, Job hears God's voice through considering
these creatures. God has spoken to him through creation, and his
relationship with God is restored.

Sadly, many churches ignore the natural world in their worship,
despite the fact that the Psalms show creation as a stepping stone into
worship. Many people today live in such urban surroundings that
they hardly hear God speaking through creation. In both cases there's
a need to recover the joy of simply spending time enjoying and
learning from God's world.

Creation: belonging to God

Psalm 24:1 declares: 'The earth is the LORD's and everything in it, the
world, and all who live in it.' This short verse is dynamite. It blows
apart the way we tend to see this world and our place within it. The
world is God's, not ours. However much we talk about 'my house',
'our country', or 'private property', none of it is actually ours. It is
all God's.

The first reason the world belongs to God is that he made it from
nothing. Psalm 24 continues: 'he founded it upon the seas and
established it upon the waters' (verse 2). If a composer writes an

original piece of music it becomes copyright and belongs to him or her. God's ownership doesn't only include the bare planet, but all the creatures he has made: 'For every animal of the forest is mine, and the cattle on a thousand hills. I know every bird in the mountains, and the creatures of the field are mine' (Psalm 50:10–11).

Over the years, Christians have sometimes taught that the earth and its minerals, plants and animals were created by God just for humans to enjoy. There is a tiny grain of truth in this, so some confusion is forgivable. In Genesis 1:29, God 'gives' plants for food. After Noah's flood, God extends this to 'everything that lives and moves' (Genesis 9:3). Later, the Psalms say: 'The highest heavens belong to the LORD but the earth he has given to man' (Psalm 115:16). People sometimes argue that God 'gave' the Promised Land, Israel, to the Jews, so it is now their possession, 'their' land.

Yet, the Bible also says clearly, 'To the LORD your God belong . . . the earth and everything in it' (Deuteronomy 10:14). How do we balance this? How can the earth be 'given to humanity' and yet also be God's? The answer is actually quite simple. Something can belong to one person in an absolute sense, yet belong to somebody else in a temporary or limited sense. If my daughter Rebekah invites her school-friends back home, she will talk about 'her' house. She's right, because it's where she lives. Yet houses don't usually belong to children, but to their parents, or, more likely these days, to the building society or bank. Rebekah has the use of it, but she is not the owner. Or imagine tenant farmers working a field. It is their field to use productively and to enjoy its fruits, but it does not actually belong to them – it belongs to the landowner.

We can see this clearly in the case of God's chosen people in their promised land. While God gives his people this fruitful land to live in and enjoy, it still belongs to him at the deepest level. God makes this absolutely clear in saying: 'The land must not be sold permanently, because the land is mine and you are but aliens and my tenants' (Leviticus 25:23). The chosen people do not, ultimately, own the Promised Land. Rather, God has given them the use of it under certain terms and conditions.

The same is true for us today. God is the landowner of planet Earth and all it contains: the atmosphere and the oceans, the mineral resources and the wildlife. Yet he has given us the use of this world as stewards and caretakers. It belongs to God and we are responsible to him for how we use and leave it. Whatever you may think of her politics, former British Prime Minister Margaret Thatcher got this absolutely right when she stated: 'No generation has a freehold on this earth. All we have is a life tenancy – with a full repairing lease.'[5] God expects us to use this world and all it contains carefully and responsibly, remembering that it's not ours but God's.

To be even more precise, the earth belongs to Jesus Christ. In Colossians 1:16 we read that 'all things were created by him and for him'. Not only does God own the world through making and caring for it, he made it *for* Jesus. The whole of creation is *for* Jesus, not for us. Why does the universe exist? It exists for Jesus. Why do the ice-caps and the rainforests, the Sahara desert and the Russian steppes exist? They exist for Jesus. What is the purpose of a blue whale, a Bengal tiger, a Pacific salmon, a house sparrow, a human being? They are all created by and for Jesus Christ, and find their purpose in him.

This has huge implications when we think about how we use the earth's 'resources'. We cannot use up the oil and gas without remembering they were made for Christ. We must take great care in experimenting with genetic modifications to species or large-scale changes to habitats, recognizing that we are answerable to God. We must not destroy forests or use up the fish in the seas – because God cares about them, and wants us to leave healthy stocks for others. We must not deliberately or carelessly allow species to become extinct because each species tells us something unique about God. Every time one becomes extinct, we are effectively rubbing out another of God's fingerprints in this world. We must treat creation with great respect, as an expression of God's character and as a possession of Jesus.

People: part of creation

Genesis chapters 1 and 2, perhaps more than any other part of the Bible, put human beings into perspective. In these ancient texts we find more wisdom than in all ancient philosophy and modern psychology combined. We are shown the nature of God, the world, and what it means to be human. As we will see, being human means understanding two essential truths about ourselves: we are part of creation, yet we are also set apart within creation.

Recently, I asked two of my daughters if they thought they were animals. Hannah, aged eleven, immediately said, 'Yes, of course we're animals; we're mammals.' Naomi, aged six, was quite disturbed by this: 'I'm not an animal; I'm a person,' she insisted. From a very early age, human beings are brought up to feel that they are different from the world around them. We see ourselves, our families, and human society as our 'world'. We believe we belong to this little human-centred world, and the bigger world outside is simply the background against which we play out the drama of our lives.

Genesis blows apart this human-centred world-view. We are not so different from the other creatures with which we share this planet. We are not descended from gods, as some religions believe. We are not specks of eternal stardust encased in human bodies, as some New Age ideas teach. We are not beings from outer space teleported onto planet earth, as science fiction dreams. Instead, we are animals (with apologies to Naomi!).

In Genesis 2:7, we are told that God formed the first human being from the 'dust of the ground'. You can't get earthier than that! Our English translations confuse us by using the name Adam, because, as somebody once said, this is really the story of Dusty and Eve! The very name Adam is from the Hebrew *adamah*, meaning 'earth' or 'soil'. Our name is 'earthling'. We are creatures of the earth, with feet of clay.

Moreover, in Genesis 1:26, God makes human beings on the same day as all the other animals. We don't even get to have our own day! We should not therefore be surprised or feel threatened if scientists tell us that we share 99% of our DNA with chimpanzees. This is not about 'creation versus evolution'. It is about our relatedness as creatures before the Creator. Probably more significant is the fact that 1% of our DNA is totally different from other great apes, roughly ten times greater than the difference between any two human beings.

Knowing our place as human beings begins by recognizing that we are part of creation. There is a sense of family likeness between humanity and every other species on earth. There is the reality that we've been created interdependent – to know and rely on each other. Until we know that we are fragile, earthly animals, we cannot really know ourselves.

People: called apart within creation

In Genesis 1:27, God says, ' "Let us make man in our image, in our likeness . . . " So God created man in his own image, in the image of God he created him; male and female he created them.'

We have focused first on the truth that we are part of creation, because this is so neglected in much Christian teaching. However, it is equally important to emphasize the other great truth in Genesis 1 and 2, that human beings are made in the image of God. The human species is a very special one, with a unique role to play in creation.

Psalm 8 wrestles with these two sides of human nature. It begins by describing the awesome power of a Creator God, and asks how it is that the One who made the whole universe can be bothered with insignificant little human beings:

> When I consider your heavens,
>   the work of your fingers,
> the moon and stars,
>   which you have set in place,
> what is man that you are mindful of him,
>   the son of man that you care for him?
> (Psalm 8:3–4)

It seems ridiculous that the Almighty Creator takes an interest in this one little species on one tiny planet in one solar system. Yet that's exactly what God has chosen to do.

God not only cares for human beings, he has also given us a role that is far more than we deserve, as the same Psalm goes on to show:

> You made him a little lower than the heavenly beings
>   and crowned him with glory and honour.
> You made him ruler over the works of your hands;
>   you put everything under his feet.
> (Psalm 8:5–6)

Humans have a place in creation that comes not by right but as a gift from God. The imagery in Psalm 8:5–6 is that of royalty: God has crowned us kings and queens, sitting on a throne with the world symbolically under our feet. Being 'under our feet' is not about trampling down, but symbolic of reflecting God's just and righteous royal authority over creation.

It is vital to hold together in tension that we are both earth-bound creatures and God's image-bearers. They are the two legs we stand on. Lose either and we will have a severely unbalanced relationship with everything around us and keep falling over.

If we forget we're made in God's image, we become only one creature among millions, nothing more than a highly evolved ape with no greater rights than any other species. Logically, this may lead us to one of two positions. The first is an amoral one, where nothing has value except in its usefulness to us. We thrive as humans simply because we're more clever and powerful. Only the strong survive, and do so by feeding off the weakest. Arguably, that's exactly how modern Western societies have treated the earth and its creatures.

Alternatively, we may take a moral stance where all species have equal value. We then run into big problems with everyday life. We have no right to use or consume other species and should all become vegans. In fact, we soon reach a position taken by the Australian philosopher Peter Singer, who has coined the term 'speciesism' to refer to the way one species (humanity) oppresses and exploits another species. Most disturbingly, if we have caused the environmental disasters facing the planet and yet have no greater significance than other species, wouldn't the planet be better off without us? The coldly logical conclusion of believing we are not in any sense special is that humanity should destroy itself for the sake of the planet. Whilst these views may seem extreme, they are attracting more and more interest today.

The opposite danger from forgetting we are made in God's image is to forget that we're made from the dust of the earth. This is a much greater danger for Christians. Environmentalists have often blamed Christianity for our current ecological crises, using Genesis 1:26–28 to argue that Christians believe humans can exploit and destroy as they wish to. The debate began in 1967 when the American scientific historian Dr Lynn White accused Christianity of being the most human-centred religion the world has ever seen.[6] If you speak to people in the green movement today, many will have accepted this view and consequently blame Christianity for the world's mess. They have a point. It is not hard to find quotations from preachers saying that the world is there for us to use and enjoy as we like. Too often churches have remained silent when the forces of destruction have been at work. Too often Christians have been so other-worldly as to be of no earthly use.

However, Lynn White and others who have developed his ideas have actually missed the point! It is not only 'Christian' countries that have caused today's disasters, but any ideology that puts human beings above nature. In 1996, I visited Siberia and went to an oil-rich area north of the city of Surgut, where I saw complete environmental devastation. Here it was atheistic communism, not Christianity, which valued only human beings and ignored the rest of creation. But there is a more fundamental way in which Lynn White was wrong. The Bible, as we've seen, teaches that creation is not for humanity but for God. Humanity is both made in God's image to care for the earth, and made from the dust as part of the earth.

So, the two sides to our human nature must be held in balance. If we ignore either, we will fail to take our true place in God's world. However, if we remember that we are both creatures of the earth and also made in God's image, this enables us to become truly human, to fulfil our God-given job description, and relate appropriately to our fellow creatures, the earth, and God.

Going back to the image of the Bible as a great drama in five acts, this first act has introduced and described the main characters: God, creation, and humanity. The plot is laid out in front of us, with a relational, caring, creative God who makes a good universe that reflects him and belongs to him, and a people who are part of that creation yet called apart to care for it. Later, we will return to what it means to be in God's image and how we work that out in practice today. For now, many questions have been answered, but a key one still remains: what will humanity do with the privilege and responsibility with which it has been entrusted?

Questions

1. What practical difference does it make to everyday life if we believe either that the earth is God, or that the earth belongs to God?
2. How should the biblical insight that 'the earth is the LORD's and everything in it' (Psalm 24:1) affect our attitude to possessions and property?
3. Why does Genesis emphasize that humans are made both 'from the dust of the earth' and also 'in the image of God'? What are the implications of each for our relationship with the rest of creation?

2 The fall: Creation's groaning

It did not take long for there to be trouble in Eden. Almost before God had finished declaring it 'very good', the harmony and beauty of creation were disturbed. Adam and Eve ate the forbidden fruit and were thrown out of the perfect garden. This dramatic event is full of important truths that still affect us today: truths about the relationships between God, his people, and the rest of creation. It's impossible to understand the state we are in, humanly and environmentally, or to understand why Jesus died, without grasping how we have fallen from God's good plans.

It is, most of all, a story of broken relationships. The friendship and intimacy that Adam and Eve enjoyed with God were gone. They had walked with God in the garden, enjoying the goodness of creation. Their nakedness showed they had nothing to hide from each other or God. After eating from the only tree God told them to avoid, all this changed. Adam and Eve hid their naked bodies from each other by making simple clothing, and when they heard God walking in the garden, they hid among the trees. The God whose purpose in creating was to enter into and facilitate loving relationships was rejected by the creatures chosen to bear his image.

The great drama of Scripture had quickly become a tragedy. Nothing would ever be the same again: a corrosive presence had entered the world and spoiled the perfection of creation. The rest of the biblical drama follows the disastrous consequences of sin's arrival and God's costly plans to resolve the crisis through Jesus. We cannot blame this on Adam and Eve alone, as the Bible is clear that every human being apart from Jesus has made the same choice. As Paul puts it, 'all have sinned and fall short of the glory of God' (Romans 3:23).

Christians have usually seen this tragedy in terms of the broken relationship between God and humanity, and rightly so. Nevertheless, its results are far wider than this. Yes, humanity falls from grace, goodness and intimacy with God. However, it is not only about one broken relationship, but about how all the good relationships that God created have been spoiled.

A simple diagram may help. As we have seen, there are three main actors in the drama of creation: God, humanity and the rest of creation. We can picture the relationships as a triangle (see figure 1).[1]

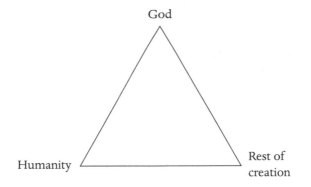

Figure 1

As human beings, we have a relationship both with God and with the rest of creation. In addition, creation itself has a relationship with God, as we saw earlier. When human beings turn against God, this not only breaks the relationship with God, but also affects the other sides of the triangle. When climbers are harnessed together by a single rope, the fall of one pulls the rope and inevitably affects all the other climbers. When the knitter of a complex pattern makes a single error, the whole item may need to be unravelled. The very relational, interdependent nature of creation means that one broken relationship affects all the others.

Because human beings turned against God's good plans, the broken relationships are seen in at least four directions, spelt out in Genesis 3 and elsewhere in the Bible:

- between God and humanity
- between human beings
- between humanity and the natural environment
- between God and his creation

Often, Christians only concentrate on the first of these broken relationships. We are thrown out of God's presence and no longer have a relationship of closeness and intimacy. From God's perspective, we, the creatures he made in his image to relate to him, are now cut off and in rebellion against him.

Yet, the impact of the fall goes further, as a damaging rift immediately opens up between the first two human beings. When God asks Adam why he has eaten the forbidden fruit, he blames Eve: 'She made me do it!' Instead of being a companion and helper, she has become a rival, even an enemy. The battle of the sexes has begun. Soon afterwards, relationships between people go further downhill, with jealousy and then murder (Cain and Abel in Genesis 4). This spills out into misunderstanding and enmity between different cultures and language groups (the tower of Babel in Genesis 11).

Sexism, racism, ethnic hatred, aggressive nationalism and all other forms of prejudice can be traced to this breakdown of relationships starting from the fall. The cause of these problems is sin: rebellion against God. Their cure is the good news (the gospel) of Christ. To repeat what Paul writes in Galatians 3:28: 'There is neither Jew nor Greek, slave nor free, male nor female, for you are all one in Christ Jesus.'

In addition to conflict between people, it is the other two areas of broken relationship that Christians have most often failed to see. The shattering effects of sin and the fall include a total breakdown in our relationship with other species and the planet, and in the relationship between God and the whole created order. In Genesis 3:17–19, God tells Adam:

> Cursed is the ground because of you;
>> through painful toil you will eat of it
>> all the days of your life.
> It will produce thorns and thistles for you,
>> and you will eat the plants of the field.
> By the sweat of your brow
>> you will eat your food
> until you return to the ground,
>> since from it you were taken;
> for dust you are
>> and to dust you shall return.

The '*ground*', the earth itself, is cursed because of the fall, as well as our relationship with the earth. Work, instead of being fruitful effort, becomes sweaty, painful hard labour. Gardening moves from being joyful partnership into a struggle to tame unruly nature. God reminds Adam in no uncertain terms that he is no more than dust, the earthling, a carbon-based life-form who will rot back into the earth from which he came.

Because of humanity's ongoing rejection of God, there is now a cancer at work within creation, causing pain and destruction from the inside. Sin has entered the human heart, causing an inner battle between good and evil like that between healthy and cancerous cells. Evil has also entered the natural world, affecting both our and God's relationships with the earth. That is why, when Jesus was born, 'He was in the world, and though the world was made through him, the world did not recognize him' (John 1:10). The curse has even affected creation's ability to recognize its own Creator. As we will see in chapter four, sin's toxic growth can only be halted when God himself chooses to intervene in costly love, through his Son's death on the cross.

Pain and suffering in a fallen world

Today we see the results of sin, and the curse which followed, all around us in creation. In an important passage in Romans 8, Paul tells us that creation is 'subjected to frustration' and 'in bondage to decay' (verses 20, 21). There is no escaping that beyond a certain age (calculated by most of us as at least five years beyond our own age!), our bodies become more and more frustrating to live in, and we notice the signs of decline: in our eyesight, our joints, and in our ability to remember small details. Like an ageing body, the whole world is a frustrating place to live in, with decay, suffering, and system malfunctions. Scientists talk about the process of entropy, where everything in the universe is gradually decaying and falling apart. As we see natural disasters on a vast scale, volcanic eruptions or tsunamis caused by undersea earthquakes, we can see the groaning of creation. No one person or group of people is responsible, yet, in a sense, we are all responsible insofar as these events are part of the curse on creation caused by human sin.

We must tread carefully, though, for there are many places in nature where pain, suffering and cruelty seem to serve a greater good. For example, leaves falling off trees, dying and rotting, provide renewed fertility for the soil. At a more complex level, when I visited Canada's beautiful west coast, I was told how conservation of endangered Pacific salmon is a vital part of maintaining not just one species, but a whole ecosystem . . . and all because the salmon suffer and die. It begins when the salmon hatch in small streams running through the evergreen forests of British Columbia. As they grow, they swim downstream, joining larger rivers and eventually finding their way out to sea into the massive Pacific Ocean. The cool waters of the deep Pacific are full of tiny nutrient-giving organisms, and the salmon grow and fatten on their rich diet. Eventually, an inbuilt instinct draws them back to their native river and stream, where the females lay eggs and the males fertilize them. By now they are often huge, but their job is done. In death they provide good food for bears, eagles, gulls and many other predators. The nutrients from the cold deep Pacific which fed the salmon are recycled through the creatures that eat them, and these nutrients pass into the soil, enriching it and causing the coastal forests of British Columbia, including huge redwood trees, to be wonderfully fertile. All of this happens because the salmon suffered and died.

In the end we are left with a mystery about how far we can attribute all 'natural' suffering and pain to the fall and sin's entry into this world. It is hard to imagine a tiger or a killer whale that is not a carnivore. Yet, although we are left with some unanswered questions, we can be clear that this earth has been negatively affected by the presence of sin, and continues to be so.

The roots of the environmental crisis

Today, more and more people are recognizing that the environmental crisis is at root a spiritual crisis. We have damaged the world not only through inadequate information or poor decisions, but through selfishness. Every time we are too lazy to walk to the local shops, and take the car, we make a spiritual choice to be selfish, and say 'no' to treating the earth as if it really is the Lord's. Whenever we buy cheap meat without asking whether the animal has been intensively and cruelly farmed, we show disrespect to God as Creator. When we know the facts about energy wastage, and the damage that excessive carbon dioxide does to the atmosphere, and yet carelessly leave the television on standby or lights on around the house, we are not only harming the planet but also sinning against God and our fellow humanity. These are uncomfortable truths, and I know that I am often the first to fail in these areas.

Yet it is vital that we realize the links between our relationship with God and our relationship with the planet. In the book of Hosea, the prophet looks around, seeing failing harvests and a collapsing ecosystem. In his analysis of what has gone wrong he doesn't blame God, nor does he blame the curse put on creation at the time of Adam. Instead he blames the behaviour of the people around him:

Hear the word of the LORD, you Israelites,
    because the LORD has a charge to bring
    against you who live in the land:
'There is no faithfulness, no love,
    no acknowledgement of God in the land.
There is only cursing, lying and murder,
    stealing and adultery;
they break all bounds,
    and bloodshed follows bloodshed.
Because of this the land mourns,
    and all who live in it waste away;

the beasts of the field and the birds of the air
  and the fish of the sea are dying.'
(Hosea 4:1–3)

Notice the clear link between cause and effect. It is because of the people's sin that the animals, birds and fish are dying. Importantly, it is not only 'environmental sin' that harms creation. Obviously creation will be harmed when we live unsustainably or over-consume: when we chop down forests without replanting or pollute the atmosphere beyond its capacity to absorb. Here, though, it is the *moral* failure of the people that has an environmental result. Lying, stealing, murdering and immorality not only have an effect on our relationships with other people and with God. They also affect the natural world. We may not be able to see the links between our moral failure and birds and animals, but they exist nonetheless. Remember the triangle of God–people–creation? As one side is broken, every side is affected. When we behave selfishly towards God, it is like breaking one strand of spider's silk and the effects rippling throughout the web.

We must be cautious here. Hosea is not claiming that every ecological disaster is caused by the people in that immediate area. We cannot say, 'If you commit adultery, you won't get rain next year'! The fact that Somalia and Sudan have suffered terrible famines does not mean that the people of North Africa are more sinful than others. Poor farmers in developing countries are often the first to suffer from extreme weather or disease, but they may be largely the innocent victims of pollution or greed elsewhere. During the 2001 foot and mouth crisis in Britain, Bishop James Jones was attacked for saying, 'I believe that the various farming crises over the years may well be a judgment of God on the way we are violating creation'.[2] The media seized on this, as if the bishop was saying that wicked farmers had brought this on themselves. Actually the bishop was simply repeating the biblical truth that we've been examining. In the end, we are all to blame. Failure to keep God's ways, whether morally or in good stewardship, has an inevitably negative effect on the land as well as on our relationship with God. A break in one side affects the whole triangle; a tear in one strand damages the whole web.

There is another reason for caution. It is very difficult to distinguish between God's judgment and the random chaos in the natural order caused by the fall. When a tower fell in Siloam killing local residents, Jesus refused to blame the individuals concerned (Luke 13:4–5). In the New Testament there is more emphasis on God's final judgment at the end of time, and centrally on the cross of Jesus, where God takes the inevitable judgment that sin requires upon himself.

We must resist the temptation to blame every farming crisis or natural disaster on specific acts of moral disobedience. On the other hand, we cannot avoid the conclusion that poor stewardship and moral decay lead to environmental disaster. 'Do not be deceived,' wrote Paul, 'God cannot be mocked. A man reaps what he sows' (Galatians 6:7). Today we are reaping what we and our forebears have sown in terms of pollution, resource depletion and climate change. The scale of crisis is larger than anything we have ever faced before. Unless we recognize that this is a spiritual crisis, we will not be able to solve it.

The scale of crisis is larger than anything we have ever faced before. Unless we recognize that this is a spiritual crisis, we will not be able to solve it.

For much of the past forty years, the environmental movement has been working to solve the growing environmental crisis. Ever-clearer scientific evidence of our effect on the planet has gradually convinced most sceptics. The world's best scientists have worked to develop new technologies to replace old polluting ones. Yet, today an increasing number are recognizing that the crisis cannot be solved by science and education alone. Professor Sir Ghillean Prance, former Director of the Royal Botanic Gardens at Kew, says, 'Science alone will not be able to resolve the situation because it is a moral, spiritual and ethical one requiring major changes in our behaviour.'[3]

Despite all the media coverage on climate change and many awareness campaigns, there has been very little change in people's lifestyles. What we need more than anything else is a profound change of heart. Individually and collectively, we have been living an impossible dream. We must admit that we have broken the sacred trust God has given us: to care for planet Earth. We need to repent, because we have sinned against God, creation and our fellow people by living selfishly and thoughtlessly.

Having preached this message for some years, both in and beyond churches, I find it encouraging that others (including secular conservation organizations) are saying something similar. The 2002 Johannesburg World Summit on Sustainable Development was a watershed. Many of those present had attended the 1992 Rio Earth Summit, where there was great optimism that with international cooperation the world's environmental problems could be addressed. Ten years on, and despite huge efforts, there had been little progress, largely because people did not want to give up their polluting lifestyles. Johannesburg saw a call for the world's faith communities to get involved, implicitly recognizing that religions know something about changing people's inner motivation. As an example, the World Wide Fund for Nature (WWF) has started a programme called 'Sacred Gifts for a Living Planet', stating in a press release that it was 'encouraging the faiths to enlarge the significant role they play in caring for the environment'.

In a paper called 'The Death of Environmentalism', Michael Shellenberger and Ted Nordhaus write: 'Environmentalists need to tap into the creative worlds of myth-making, even religion, not to better sell narrow and technical policy proposals, but rather to figure out who we are and who we need to be.'[4] The spiritual crisis that has affected planet Earth is finally affecting the green movement, as environmentalists seek a way of seeing the world that explains the damage we have done and gives hope for the future. It is a time of great opportunity for Christians, if only we will look again at the Bible, listen to God's challenge, and be prepared to change our own lifestyles.

Judgment and hope

How does God feel about the spiritual rebellion that has led to such devastation in his beautiful creation? The last book in the Bible tells us straight: 'The time has come for judging the dead ... and for destroying those who destroy the earth' (Revelation 11:18). In another translation it simply says: 'God says, "I will destroy those who destroy the earth."' Does God care when our culture pours noxious gases into the air, pollutes the seas, fills the land with waste, and treats the earth like an unwanted plaything? Of course he does, and his anger is building against those who destroy that which he declared 'very good'.

Nevertheless, alongside God's judgment comes God's mercy. Romans 8 moves from creation's groaning ('subjected to frustration' and in 'bondage to decay', verses 20–21), to the beginnings of new hope beyond judgment. Creation's groaning is not the death-throes of a dying world, but is described as 'the pains of childbirth' (verse 22). Childbirth is not about despair but the 'hope that the creation itself will be liberated from its bondage to decay and brought into the freedom of the children of God' (verse 21). To see exactly what that means we will have to wait for chapter five, but let us conclude with this thought.

Despite all that has gone wrong with this world, God has not abandoned his creation. God's image may be distorted and the seeds of ecological disaster may be sown, but the Psalms tell us that the earth still expresses God's power, glory, character and provision. The mountains, the seas, the rivers, the animals and birds are still the mighty works of God (Psalms 19 and 147). However damaged God's works are by human folly, they still witness to this great God, and God still sustains his creation.

Questions

1. What examples can we see in the world today, and in our own lives, of the breakdown in relationships between humanity and creation?
2. How far can we attribute all suffering and pain to the fall and sin's entry into this world? Think of some actual examples from recent news or your own experience.
3. The World Wide Fund for Nature writes of encouraging faith communities to 'enlarge the significant role they play in caring for the environment'. Are there ways you could engage in this process?

3 Land: People and place in context

'Land' is perhaps the most overlooked theme in the Bible. When people think of the Christian message, they don't immediately think of 'land'. In summing up the Christian gospel, preachers usually jump from the problem (the fall), straight to the solution (Jesus).

Yet most of the Bible lies in between the fall and Jesus. This is the third and central act of the biblical five-act drama. It lasts from the expulsion of Adam and Eve from Eden, until the coming of Jesus. The key theme of this act is Israel – the history of a people and a land. More than anything, it is the tale of the relationship between the 'chosen people' of Israel and the 'promised land'. It is not in the Bible just as padding to keep us hanging in suspense until we get to the arrival of Jesus, but to give us vital insights into how we are to live as real people in a fallen world.

The Old Testament refers to land 2,000 times, and the New Testament 250.

To give us some idea of how important and overlooked the theme of 'land' is, it's worth noting that the Old Testament refers to land 2,000 times, and the New Testament 250.[1] In the life of Abraham alone, forty of the forty-six promises God makes mention the land and twenty-nine are mainly or exclusively about land. It's been calculated that there are more references in the Bible to land than to justification by faith, repentance, baptism or Christ's return.[2] In his important book, *The Land*, Walter Brueggemann argues that 'land is a central, if not the central theme of biblical faith'.[3] However, if you go to church, when did you last hear a sermon on 'land', beyond a vague harvest festival talk? In an era when changing land-use, scientific dilemmas, resource depletion and global ecological crises threaten life itself, perhaps God wants us to rediscover just how to treat the land we inhabit.

The land belongs to God

In act 1 we saw that the whole creation belongs to God (Psalm 24:1). Thus, although we can talk about 'my land' or 'our country', and people may buy and sell land, human landownership is always secondary to God's. Human beings can only ever be lease-holders or managers of God's land. As Crocodile Dundee famously said in the (first) film of the same name, 'For people to argue about who owns the land they live in is a bit like two fleas arguing about who owns the dog they're on!'

When God starts making promises to Abraham about giving him 'a people' and 'a land', what does he mean? These words come soon after humanity's failure to fulfil God's first plan: to display his image through caring for the land and all that is upon it. In calling Abraham and promising him as many descendants as the stars in the sky, God is beginning his rescue plan. If humanity as a whole has abandoned God's perfect will, then at least there could be a chosen people, a holy nation set apart and bound to God by agreeing to keep his promises. The destiny of this chosen people is to occupy a special land, a place where they can fulfil the human destiny of imaging God in creation care. The destiny of Israel is not primarily a political one. This isn't mainly about 'land' as a state with high walls and border guards. It is more about land in a spiritual and ecological sense, as God's chosen people are to model a godly relationship between humanity and the earth by living in the land.

The relationship between the people (of Israel) and the land (of Israel) begins with a promise. The land is God's to give, and it is God who promises that Abraham's descendants will occupy it. Even though the people conquer the land by force, it is made clear that this is only because God gives them the land. Even when they are safely settled, the land still belongs to God, not to the people. In Leviticus 25, the Lord commands Israel to observe a rest, a sabbath, on the land every seven years, and a jubilee every fifty, when sold land would be returned to the original owner's family. God reminds the people: 'The land is not to be sold permanently, because the land is mine and you are but aliens and my tenants' (Leviticus 25:23).

As an expression of the fact that the land remained God's, Old Testament landownership was radically different from that of neighbouring societies. In other cultures, kings and powerful chiefs owned most of the land. However, in Israel, 'the land was divided up as widely as possible into multiple ownership by extended families. In order to preserve this system, it could not be bought and sold commercially, but had to be retained within kinship groups'.[4] The story of Naboth's vineyard in 1 Kings 21 illustrates this. King Ahab desired Naboth's land and tried to buy it, but Naboth replied, 'The LORD forbid that I should give you the inheritance of my fathers' (1 Kings 21:3). Naboth did not see the land as his to dispose of. He was tied to the land: bound to his ancestors who had occupied it and his descendants to whom it was promised. Together they were bound to God as the ultimate landowner.

We belong to the land

We have looked at the triangle of relationships between God, people and creation. In a similar way, much of the Old Testament is about the relationships between a holy Creator, a chosen people and a promised land (see figure 2). Just as the people are connected in relationship to God, and as God is connected to and owns the land, so people are themselves tied to the land.

The relationship between human beings and the land goes back to creation. Being made from the dust of the earth (Genesis 2:7) gives us

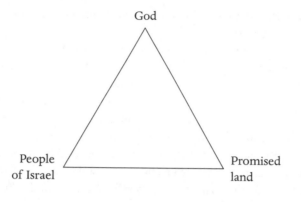

Figure 2

a kinship with the soil itself. We have already seen that in Genesis 2 the Hebrew word for 'man' (*adam*) is deliberately derived from the word for 'ground' (*adamah*). The words 'human', 'humus' (soil) and 'humility' share a common root. We should have a true humility in terms of our place on this planet.

God has not created us as disembodied souls, living in a vacuum. Rather he has made us to live in a physical, geographical, earthy context. Just as God is a relational God, so he has made us relational beings, tied to the earth by our dusty origins. As physical beings, we need to be 'earthed', 'rooted' and 'grounded', or we become 'dislocated', 'uprooted' and 'displaced'. Russ Parker says: 'One of the strongest and most basic needs of the whole human race is to belong, and to belong in the place or on the land where we can connect, be rooted and grow.'[5]

Those who live in large cities, with rapid transport from one man-made environment to another, cut off from the cycles of the seasons and changes in the weather, lose much of their connection with place. Most people no longer live in one familiar place for most of their lives, with their families and community around them. And yet, even in this postmodern globalized world, the biblical story can give us new hope.

The people of Israel knew both about wandering and being settled. They spent time as nomads and exiles, and also as settlers in the land they were promised. In Jeremiah 29, we read some words spoken to a time of dislocation and uprootedness such as ours. They are God's words to a people in exile, carried into a country that was not their own:

> This is what the LORD Almighty, the God of Israel, says to all those I carried into exile from Jerusalem to Babylon: 'Build houses and settle down; plant gardens and eat what they produce. Marry and have sons and daughters; . . . Also, seek the peace and prosperity of the city to which I have carried you into exile. Pray to the LORD for it, because if it prospers, you too will prosper.'
> (Jeremiah 29:4–7)

I live in a part of London that is a mosaic of many cultures and languages, where Jeremiah's words have a particular resonance. Around me are people who have come to Britain from Sri Lanka, Iraq, Afghanistan, Somalia, Pakistan, Grenada and India, amongst many other countries. Home is a fond memory from far away and long ago, and urban London is a cold, dirty and miserable place. Like the people of Israel in exile in Babylon (to whom Jeremiah 29 is addressed), many wish they could be somewhere else and have no desire to put down roots.

Yet God's word is clear: 'Build houses and settle down; plant gardens and eat what they produce . . . seek the peace and prosperity of the city to which I have carried you into exile.' God wants us to put down roots wherever we are, even if it is not the place we have chosen or even like. These roots are essential for our psychological and spiritual health as human beings. They are also essential if we are to have a healthy relationship with the land: the natural environment around us.

The importance of being rooted in a particular place continues into the New Testament. Jesus modelled a relationship with place, spending thirty years in Nazareth as part of a local community closely connected to the land. I sometimes speak of the 30:3 principle. Whereas today we might expect somebody to train for three years before working professionally for thirty, Jesus did the opposite. He spent thirty years putting down roots into a particular culture, place and environment, before moving to a public ministry that only lasted for three years.

Later on, speaking in Athens, Paul explains how the Creator God 'determined the exact places' where people should live (Acts 17:26) *so that* they would seek and find him (verse 27). It is in our long-term interaction with creation in a particular place that we find hints of God's glory and begin to seek for a relationship with God.

Christians have often over-spiritualized faith, forgetting that it is closely connected with place. This world is our God-given home. It is a good home, although much spoiled. It contains wonderful diversity and beauty and is blessed by God's Son Jesus making it his home too. Just as Jesus worked out his relationship with God in the back streets of Nazareth, the roughness of a carpenter's shop and the deserts of Sinai, so we are to put down roots wherever God has planted us. We will see later that this biblical theme has practical everyday implications for how we live our lives.

Land as context for our relationship with God

For many people today, Leviticus, Numbers and Deuteronomy are amongst the hardest books in the Bible. They seem to be full of detailed regulations with very little relevance to modern life. However, once we understand the importance of land, some of these passages suddenly make more sense.

As God's image-bearers, God gives us detailed instructions about looking after the land and its creatures. As Dr Chris Wright states, 'Nothing that you can do in, on or with the land is outside the sphere of God's moral inspection.'[6] The detail of Old Testament law touches on every detail of farming practice, from the welfare of domestic animals to the need for margins at field edges.

For example, in Deuteronomy 22:6–7 we read: 'If you come across a bird's nest beside the road, either in a tree or on the ground, and the mother is sitting on the young or on the eggs, do not take the mother with the young. You may take the young, but be sure to let the mother go, so that it may go well with you and you may have a long life.' These verses, which may at first sight appear random or irrelevant, are about our relationship with creation and with God and give us important, practical principles for today. God allows his people to 'harvest' wildlife for food – to take eggs or young birds – but they must leave the mother bird alone, so that she can lay more eggs and raise a second brood. Here is a principle of sustainable living for us to re-learn.

Not just with bird's nests but with every aspect of God's creation we should ask not only, 'What do I get out of it in the short-term?', but also 'What will the long-term effect of my actions be on the natural environment?' For instance, if only we applied this principle of sustainable use to the oceans, we would not have tragically over-fished to the extent that some species may never recover and the fishing industry is now in crisis. Living sustainably is about using creation's resources only in a way that allows them to replenish themselves.

In these verses, God also makes it clear that our own welfare is tied to respecting and preserving wildlife. The result of letting the mother bird go is 'that it may go well with you and you may have a long life'. Just as the people of Israel's well-being in the land was tied to obeying God's advice in their relationship with the land, so our well-being and even survival depend on rediscovering our inter-dependence within God's creation.

In another passage, Leviticus 19:9–10, we find these words: 'When you reap the harvest of your land, do not reap to the very edges of your field or gather the gleanings of your harvest. Do not go over your vineyard a second time or pick up the grapes that have fallen. Leave them for the poor and the alien. I am the LORD your God.' Moderation is the important principle here. In the past sixty years, all over the world, there has been huge pressure on farming land to be more productive. This has been achieved through artificial fertilisers and pesticides, by improved crop varieties, and by squeezing results from every marginal corner of farmland. Sometimes farmers have been guilty of greed, but more often they have been the victims of pressures from government policies, supermarkets or global economic forces.

This passage shows us that land-use should not be about productivity alone. The size of our harvests and our profits is not the only thing that matters to God. Leaving the 'gleanings' – the grain that has fallen or has grown at the margins – provides for the poor, the foreigner and, of course, for wildlife. Super-sized mono-culture fields with no hedges or margins may make big profits, but they are disastrous for wildlife and ultimately dishonour God.

These instructions about leaving enough for the needy are not simply social or even ecological. They are also spiritual, for 'I am the LORD your God'. These last six words of Leviticus 19:10 should be inscribed in the consciousness of every one of us as consumers. It's God's world, and he will ask us to give account of our stewardship of its resources.

One final principle is that of sabbath for the land. In Leviticus 25:2–7, God tells the people: 'When you enter the land I am going to give you, the land itself must observe a sabbath to the LORD. For six years sow your fields, and for six years prune your vineyards and gather their crops. But in the seventh year the land is to have a sabbath of rest, a sabbath to the LORD. Do not sow your fields or prune your vineyards. Do not reap what grows of itself or harvest the grapes of your untended vines. The land is to have a year of rest. Whatever the land yields during the sabbath year will be food for you – for yourself, your manservant and maidservant, and the hired worker and temporary resident who live among you, as well as for your livestock and the wild animals in your land. Whatever the land produces may be eaten.'

Here, written thousands of years ago, is a clear understanding that land cannot constantly be farmed intensively, but needs rest. When the European Union introduced the principle of 'set aside' land – paying farmers to leave certain fields unfarmed, they didn't realize they were reinstating an imperfect version of the biblical sabbath principle!

The principle of rest is written into the very order of creation, part of the seven-day structure of Genesis 1 and 2. Not only people, but animals and the land need rest, and all suffer without it. What is most striking is the way this passage ties the relationship between people and land to relationship with God. This is a sabbath 'to the LORD'. We cannot have a completely healthy relationship with God unless we treat the land in a godly way. Through celebrating sabbath rest we can also recover a sense of dependence on God. The land produces its fruit not because of our hard work, but because of God's provision and care even while we rest!

All these Old Testament laws were given for Israel's particular situation and should not necessarily be transposed literally; but they provide underlying principles that still apply today. God takes a lively interest in the welfare of all his creatures and the land itself, and humans are responsible to God for how the land is treated. The principles are taken from creation itself and thus have universal relevance.

The land acts as a spiritual barometer

The Old Testament paints the constant ups and downs of Israel's relationship with God and the resulting effects on creation. Times of rich harvest and bounty, or of famine and exile, are all linked to spiritual obedience and disobedience. As Chris Wright says of Deuteronomy 28 – 30: 'The land itself will be both the arena and agent of God's blessing or curse.'[7]

In the great biblical drama, the land is not just scenery or backdrop, but an active character in the story. Time and again, the Bible uses active verbs, not to suggest that the land itself is alive, but to emphasize that it is the active agent of God's blessing and curse. It is worth looking at a few examples.

In the Psalms, we read:

Shout for joy to the LORD, all the earth,
    burst into jubilant song with music . . .
Let the sea resound, and everything in it,
    the world and all who live in it.
Let the rivers clap their hands,
    let the mountains sing together for joy . . .
(Psalm 98:4, 7–8)

Even inanimate objects – rivers, trees, mountains and soil – are invited to praise God. When God acts, the land reacts. Psalm 66 invites the land to praise God for his actions:

Shout with joy to God, all the earth!
    Sing the glory of his name;
    make his praise glorious!
Say to God, 'How awesome are your deeds!
    So great is your power
    that your enemies cringe before you.

All the earth bows down to you;
　they sing praise to you,
　they sing praise to your name.'
(Psalm 66:1–4)

However, not only does the land react positively to God's gracious actions; it also reacts negatively when people fail to keep God's commands, and when God is obliged to send his judgment upon creation. We have already examined Hosea 4:1–3 and the links between our moral failure and the ecological problems we see around us. This is a theme that the prophet Jeremiah also comments on:

How long will the land lie parched
　and the grass in every field be withered?
Because those who live in it are wicked,
　the animals and birds have perished.
Moreover, the people are saying,
　'He will not see what happens to us.'
(Jeremiah 12:4)

It is clear that the land is acting as a spiritual barometer – reacting to the moral disobedience of God's people by 'mourning' and wasting away. Wildlife and natural systems are all affected. As in Hosea, the cause is not simply poor stewardship, or ecological mismanagement. It is moral failure – cursing, lying, murder, stealing and adultery – which have a direct result in environmental disaster. Elsewhere, sins such as idolatry (Jeremiah 3:6–10; 16:18), bloodshed (Numbers 35:33–34) and broken promises (Isaiah 24:5–6) all have their effect on the land.

The image of the triangle of relationships helps explain the link between moral failure and ecological damage. The line linking God and humanity is broken when we sin against God, and creates a knock-on effect on the other two sides of the triangle.

When God's people keep damaging their land by failing to bear his image in how they treat it, they eventually get found out. Just as a stomach violently expels a foreign body, so the land forcibly ejected the people of Israel into exile: 'And if you defile the land, it will vomit you out as it vomited out the nations that were before you' (Leviticus 18:28).

Today, we stand at a time of increasing ecological disaster. Some is simple cause and effect. Poor stewardship, bad management and greed have a direct effect. However, some of this may also be our reaping what we have sown in terms of moral disobedience. We must be careful not to blame every natural disaster on moral failure by local communities: the results of sin are distributed unevenly and often randomly across creation. Still, the underlying biblical principle is inescapable: failure to keep to God's ways has an inevitably negative effect on the land, as well as on our relationship with God.

Healing the land?

If the land suffers from the results of human sin, what can be done to heal it? The phrase 'healing the land' occurs only once in the Bible, in 2 Chronicles 7. The setting is King Solomon dedicating the new temple building in Jerusalem to God. Sacrifices are made, God's glory comes down, filling the temple, and then God himself speaks to Solomon:

> 'I have heard your prayer and have chosen this place for myself as a temple for sacrifices. When I shut up the heavens so that there is no rain, or command locusts to devour the land or send a plague among my people, if my people, who are called by my name, will humble themselves and pray and seek my face and turn from their wicked ways, then will I hear from heaven and will forgive their sin and will heal their land.'
> (2 Chronicles 7:12–14)

'Healing the land' has become a popular phrase in worship songs and amongst preachers. People use it to speak of God's favour returning on a nation that has turned against him. However, this is missing the point! 2 Chronicles 7:14 is talking about 'land' in an ecological rather than a social or political sense. This is not primarily about healing a nation, but about healing a place – a physical geographical region – from problems such as drought, disease and crop-failure, all mentioned in verse 13.

What is so significant is that healing the environment comes about not primarily by recycling, down-sizing or resource management, but by repentance and returning to God. The land can only be healed when its inhabitants recognize whose land it is, and repair their broken relationship with God and each other. If the ecological crisis is ultimately a spiritual crisis, then the cure is also a spiritual one.

We must be careful, because there is a potential danger of what could be termed an 'eco-prosperity gospel', where bumper crops are a sign of spiritual purity. God is not a machine to be manipulated by our prayers or our righteousness. Bumper crops are not always a sign of God's blessing. They may simply be signs of good weather, or even the result of agricultural policies which achieve short-term gain but damage the long-term health of the soil and the whole ecosystem.

There have been extraordinary stories around the world of how the land can be transformed when a community turns to God in repentant prayer and renewed obedience.

The Bible does not promise material prosperity or physical health in this life, nor does it promise a completely restored environment now. The entry of sin into the world has led to randomness in how nature operates. It is the whole creation, not just sinful people, that has been 'subjected to frustration' and is 'groaning as in the pains of childbirth' (Romans 8:18–25) because of human sin. An undersea eruption in the Pacific may cause flooding in Japan, a nuclear disaster in the Ukraine may lead to irradiated sheep in Cumbria, industrial pollution from Britain may damage forests in Scandinavia. Globally we are all in this together.

There is also another, opposite danger: thinking that faith and land have no connection. If we find it hard to imagine how repenting and returning to God can benefit the natural environment, it's because we've forgotten how everything is connected in God's world. The promise of 2 Chronicles 7:14, while given to Solomon and the people of Israel, has a more general application for us today. God longs to bring healing to the land, and as people repent and return to him, both in personal morality and in their stewardship of the earth, God's healing grace is released into healing the land itself.

There have been extraordinary stories around the world of how the land can be transformed when a community turns to God in repentant prayer and renewed obedience. One of the most remarkable has been reported from Almolonga in Guatemala, where almost sterile land has been transformed into fertile fields yielding huge vegetables as a local community has turned from crime and immorality to Christ.[8] Such stories may be extraordinary, but if we believe that God is Creator and take the Bible's message seriously, this is exactly what we should expect.

If we wish to see the land healed, we should examine our own lifestyles and our use of God's resources. At a broader level, we need to look at farming and environmental policies in recognition that all things belong to God and we are answerable to him for how we use his world. Furthermore, if our relationship with the land is to be healed, we must ensure that our relationship with God is healed. We can bring to him any ways in which we are failing him and ask for his forgiveness, and for a restored relationship. All of this is only possible because of Jesus, and we turn now to look at his work in restoring all our broken relationships.

Questions

1. 'God wants us to put down roots wherever we are, even if it is not the place we have chosen, or even like.' How deep are your roots into your local community and environment? Using Jeremiah 29:4–7, think about how they could grow deeper.
2. How far is it right to speak of 'natural' disasters or farming crises as God's judgment upon our sinful behaviour?
3. In 2 Chronicles 7:12–14, what does God require of the people before he will 'heal' their land? How can we translate this message into today's context of climate change and environmental uncertainty?

4 Jesus:
Saviour of the world

Recently I attended the memorial service of a man who had died quite suddenly in his fifties. I'd known him for a short time and only knew a few things about him: his Christian faith, his commitment to his family, his love and knowledge of nature. At the service I discovered a much more complete picture as people from different areas of his life paid tribute to their friend and colleague. I found he had lived in the same city all his life, through school, university, work, marriage, church, quietly committed to the place and its people. I learned he had a PhD in botany, something he had never boasted about or publicized, but which helped explain his excellent knowledge of natural history.

I heard how he and his family had moved from a large city-centre church to a small under-resourced church on their local housing estate and had thrown themselves into helping in practical ways, building deep and lasting relationships. When a work colleague stood up and read out dozens of short tributes, I discovered a whole new side to him. He had worked for the local council for over thirty years and had patiently encouraged many others. Several significant city-wide environmental initiatives had been inspired by him, and his gentle wisdom and advice had affected the whole council staff, many of whom attended the service. I thought I had known this man fairly well, but I discovered I'd only known a tiny fraction of who he really was.

The same is often true of Jesus. To give two examples, the people of Nazareth knew Jesus as the carpenter, skilled with woodwork, a loyal son to Mary after Joseph died. As a result, they struggled to relate to him when he returned to Nazareth with a band of followers, claiming to speak with God's authority. They tried to throw him over the cliff of the hill on which Nazareth was built, and Jesus had to leave. Much later on, in the Middle Ages, most Christians could not read the Bible and lived in societies where the church was closely tied to the state. The images they had of Jesus were the ones painted on church walls, often of a triumphant powerful King on a throne, ruling over the world. Both the people of Nazareth and those in the Middle Ages had a valid picture of Jesus, but both were incomplete. Today, as twenty-first-century Western Christians, we often focus on a very personal side to Jesus: 'My Jesus, my Saviour', a friend we can turn to at any time. Again, this is helpful and accurate, but it isn't the whole picture.

If we think that the life and work of Jesus had very little to do with the environment, perhaps this says more about us than about the Bible. As I've re-read what the Bible says about Jesus, I am astonished at how much there is of relevance to the earth, and our relationship with, and God's purposes for, creation.

We have already seen how Jesus is vital in understanding the story of creation. Colossians 1:15–20 is a passage that puts the earthly life of Jesus into a perspective that is both cosmic and eternal. As you read these words, try and hold them together with images of the man who walked this earth for thirty-three years, as well as the Lord to whom Christians pray today.

He is the image of the invisible God, the firstborn over all creation. For by him all things were created; things in heaven and on earth, visible and invisible, whether thrones or powers or rulers or authorities; all things were created by him and for him. He is before all things, and in him all things hold together. And he is the head of the body, the church; he is the beginning and the firstborn from

among the dead, so that in everything he might have the supremacy. For God was pleased to have all his fullness dwell in him, and through him to reconcile to himself all things, whether things on earth or things in heaven, by making peace through his blood, shed on the cross.
(Colossians 1:15–20)

The first thing that should strike us is the scope of who Jesus is. These words blow apart any cosy, comfortable pocket-size Jesus. Instead, Jesus is:

- The *source* of creation (verse 16: 'By him all things were created')
- The *sustainer* of creation (verse 17: 'In him all things hold together')
- The *saviour* of creation (verse 20: he is the one through whom God is pleased 'to reconcile to himself all things, whether things on earth or things in heaven, by making peace through his blood, shed on the cross')

These are amazing claims. The biblical drama that this book attempts to explain hinges on these claims. There is no hope amidst the environmental crises we face apart from Jesus Christ.

We will now examine these three key claims in the light of Jesus' life, death and resurrection.

Jesus: creation's source

In act 1, when God creates the heavens and the earth, he does it by speaking creation into being. God *says*, 'Let there be light', and light appears at his command. At the start of John's Gospel, Jesus is called God's 'Word':

In the beginning was the Word, and the word was with God, and the word was God. He was with God in the beginning. Through him all things were made; without him nothing was made that has been made.
(John 1:1–3)

John wants us to know that the tiny, vulnerable, dependent baby born in Bethlehem is also Almighty God, the One through whom the whole universe was created. Jesus existed before the creation of the world. In Colossians 1:15, he is described as 'firstborn over all creation'. So, while it is right to talk and think of Jesus as a man because he became one hundred per cent human, we mustn't forget that he is also far more than that. He is *over* all creation: above it, superior to it, beyond it.

This makes the birth of Jesus as a tiny baby something quite amazing. God, the creative genius who dreamed up the whole universe, actually becomes part of creation. It's a bit like a playwright acting in their own play, or an artist stepping into their own painting, inadequate though these images are.

Knowing that Jesus is the source of creation also helps us understand some of his actions. When Jesus heals the sick, he isn't simply a human being with a magic touch or a channel for God's power. This is God himself, the Creator, fixing something that has gone wrong with his creation. If a computer engineer creates a special piece of hardware for me, where should I turn if it starts to malfunction? I could try to fix it myself, but I would only make it worse. I could take it along to a computer shop, but they would probably scratch their heads and say, 'Sorry, mate: never seen one of these before.' The one who made it is the one who knows how to fix it, and so I should take it back to the designer.

It is not just human bodies that are sick in our world. The whole creation is in need of healing. So, who's going to make it possible for things to be put right? The One who made it! There are hints of this throughout the Gospels. Jesus not only heals human illnesses but shows himself to be Lord over the forces of nature. In Luke 8:22–25, Jesus is caught up in a terrible storm on Lake Galilee. His followers, including experienced fishermen, are terrified and afraid they'll drown. Jesus is asleep and, when woken, isn't fazed by the storm. He takes control of the situation and quietly commands the storm-force winds and enormous waves to be still. And they obey him. No wonder the disciples were afraid! What power could possibly calm down the forces of nature? To the Jewish people, the sea represented the forces of chaos in the whole created order, and nobody except God himself could control them. Jesus, who had spoken these very forces of nature into being, was – and is – uniquely able to do something about them.

As well as showing Jesus' power over creation, the Gospels show us Jesus' intimacy with creation. We often miss this because we're not looking for it. But it should not surprise us – artists know their work better than anybody else! A sculptor who has worked a piece of clay into shape has a deep knowledge and relationship with her creation. In turn, an artist's work can tell us a lot about him. So it is with Jesus and the earth.

Jesus taught by stories (parables) and his teaching shows that he knew the details, patterns and rhythms of creation inside out. He taught about God by asking people to look at creation: God's book of works. His textbooks were seeds and harvests, birds and flowers, fig trees and vines, bread and wine. These weren't only visual aids to reinforce spiritual points. We miss the parables' real power if we over-spiritualize them. The elements of creation that Jesus talked about were things that he had created himself. They were, as with all creation, filled with the fingerprints of God.

It is clear that Jesus expected his followers to be familiar with the natural world. In Matthew 6, when he told people to 'look at the birds of the air' and at the 'lilies of the field', he was talking about learning from creation. The words he used were forceful: 'look and learn', 'earnestly consider', 'go and study'. Jesus was telling us all to become bird-watchers and botanists! Back in Genesis 2 God sent the first humans into the garden to give names to all the creatures. Naming is about understanding, describing and knowing. The foundation of all the biological sciences is separating creatures into their species and families through careful observation (the discipline of taxonomy). All through Scripture, we find the prophets knew their wildlife well. For instance, in Jeremiah 8:7, we find the prophet watching migrating storks and thrushes, which instinctively followed the same paths through the land each spring and autumn, and marvelling at them – and asking why God's people couldn't follow him so well!

Jesus: creation's sustainer

If a building were on fire, most of us would grab our own things first. This isn't simply selfishness; we value and protect what we are responsible for. According to Colossians 1:16, the whole of creation was not just created by Jesus, but made *for* Jesus. That three-letter word 'for' is very encouraging because, like us, Jesus cares for and protects what belongs to him. He is not only the Creator but the Sustainer of the world. God is not an absentee landlord who has abandoned the earth to be ruined by its inhabitants. Rather, Jesus is an active gardener, working through his Spirit to keep and preserve the earth, its systems and its creatures. At a time when humanity is failing to look after the world in a sustainable way, we have the assurance that ultimately Jesus is the Sustainer. He is committed to creation and will not abandon it. This does not mean human beings can sit back and leave it all to Jesus. Rather, God has chosen to work in partnership with us. He may be the Sustainer, but we are to be his stewards, fellow-workers in serving and preserving the earth.

That short word 'for' in Colossians 1:16 has another vital implication. Questions about the meaning and purpose of life have troubled people throughout the centuries. What is the world for? Why are we here? The Bible's radical and surprising claim is that Jesus is the purpose of creation. The New Testament claims that the whole of creation was made for Jesus. In a real and profound sense, the whole of creation is God's love-gift to his Son, Jesus Christ.

No wonder the universe is designed with such beauty and harmony. It is an expression of the love in the heart of God. This is why it shows so much of God's character and goodness. When we look at a sunset or a mountain range, a spider's web, a forest of autumn leaves or a new-born baby, and find our hearts lifted, we are reacting to the love of God that is at the centre of creation. As humans, we are probably the only species that can consciously ask the question, 'Why am I here?' The answer: we are here to worship God in Jesus, and to live in relationship with him within his amazing creation.

In fact, Colossians goes even further than claiming all things are made by and for Jesus. In the following verse, we read that 'in him all things hold together' (Colossians 1:17). Jesus, who made himself nothing by being born as a tiny baby, is at the heart of the whole universe. It holds together in him. Some years ago I was given a wooden toy which formed a sphere, but was made up of about eight very differently shaped wooden pieces. When all the bits were separate, I found it almost impossible to put it back together again. However, I eventually discovered that there was one key piece with which to start. As long as I built all the other pieces around this central piece in the correct order, I could eventually rebuild the wooden sphere.

Jesus is the 'key piece' in the puzzle of creation. He is at the heart of all things. They were made by him, they exist for him, and he sustains them by his power. Without Jesus at the centre of creation, the forces that hold back chaos would be removed. Scientists are discovering that this planet is incredibly finely balanced in having exactly the right conditions for life. Without trees, there wouldn't be enough oxygen; without animals, there wouldn't be enough carbon dioxide. The water-cycle, the temperature, the ocean currents, and the thickness of the upper atmosphere are all exactly right for life. What is even more amazing is that, despite major changes in climatic conditions during the earth's existence, this fragile balance has been preserved.

As they've sought to explain this, some have started talking about the earth as if it were a living organism. Professor James Lovelock's 'Gaia' hypothesis claims that the 'biosphere' (the earth as a whole, including soil, atmosphere and all the plants and creatures upon it) balances and regulates itself in such a way that it can be analysed as a single organism. Although Lovelock is not religious, many New Age thinkers have taken up this idea and worship 'Gaia' or 'Mother Earth', who looks after us so well.

If only those who talk of Gaia had looked at Colossians 1:17 in depth! The earth behaves like this, not because it is looking after itself, but because Jesus – in whom all things hold together – is looking after it. The earth cannot sustain itself but is being sustained by the One for whom it was made. God's Holy Spirit is active within creation today: sustaining and renewing the earth, providing for animals, overseeing the rhythms and patterns of the seasons, of night and day.

At a time when so much is going wrong with planet Earth, we can take comfort from this truth. Amidst worrying predictions about climate change, we need not despair, because Jesus is holding everything together. However, Colossians 1:17 also holds a hidden warning. If all things hold together in Jesus, then the converse is also

implied. Without Jesus, all things fall apart. God has entrusted
human beings with creation's care. When we fail to put Jesus at
the centre of our thought and behaviour, is it any wonder that the
environment falls apart? If we treat Earth's resources (oil, gas,
wildlife, clean air, good soil and water) as if they exist for us and
forget they were made for Jesus, should we be surprised if things
go wrong?

In our culture, and even in our behaviour as Christians, we have
removed Jesus from his position at the heart of creation. We need to
recover a sense that creation is infused with God's presence and love,
and should be treated with care and reverence because it is made for
Jesus. As I've reflected on this, I now
try to put it into practice in the detail
of everyday life. The way I use the
world's resources should reflect my
worship of Jesus: I no longer throw
away rubbish without sorting it and
recycling what I can. God made a
world where nothing is wasted, and
it dishonours him if I am careless
with what he has made. In terms of
the food I eat, I cannot in good
conscience eat meat from animals
that have suffered cruelty simply in
order to keep food prices cheap.
These are creatures which were
made for Jesus, declared 'good' by God, and entrusted to human
beings to be ruled over in a gentle and godly way. How can we say
we are worshipping Jesus if we deliberately or carelessly cause
unnecessary cruelty to his creatures? In terms of how I travel, I
now walk, cycle and use public transport more than ever before,
finding not only that I am causing less pollution, but that I have more
time to spend with God, enjoying Jesus' companionship as I travel.

> God made a world
> where nothing is wasted,
> and it dishonours him
> if I am careless with
> what he has made.

Jesus: creation's Saviour

This section is critical to the whole book. Let me sum it up in a
sentence, before explaining it: the world was created good, has been

spoilt by sin, but through Jesus there is the hope of salvation both for people and for the whole creation. For many years I didn't understand this. I believed that Jesus came to bring salvation for people and that was the end of it. The world didn't matter, ultimately because Jesus would rescue us from it. Now, I've come to see that this is only half the story. God is much bigger than I'd realized, and his purposes in Jesus are much more far-reaching than I'd ever dreamed.

What has persuaded me of this is one thing: the Bible. I've come to realize that I, along with many others, had been reading the Bible through tinted spectacles. Anything to do with the non-human creation was filtered out and ignored. Let me illustrate this using what is generally accepted as the best-known verse in the Bible:

> God so loved the world, that he gave his one and only Son, that whoever believes in him would not perish but have eternal life. (John 3:16)

I had heard John 3:16 read hundreds of times. I heard sermons about it, and I even preached from it myself without seeing what was right in front of my eyes. I always read it as 'God so loved the *people*', and it was only when I studied New Testament Greek at Bible college that it dawned on me that it actually says, 'God so loved the *kosmos*.' *Kosmos* is of course the root word from which we get 'cosmos' and 'cosmic', and its usual meaning is the whole universe. I want to be cautious here because biblical scholars are uncertain whether in this particular passage *kosmos* could mean the 'world of fallen humanity' rather than the whole material creation. So, without putting too much weight on one verse, let it illustrate that we often read the Bible and only see what we expect to see.

Our human-centred culture makes us overlook a mass of important material about God's relationship with, and saving care for, the whole of creation. The words on the page may say 'world' but we only see 'people'. However, God's perspective is much bigger than ours. He really does care about the whole of creation, and the saving work of Jesus has implications that are quite literally cosmic.

To understand Jesus as creation's Saviour, I want to begin in what may seem a surprising place: the account of Noah in Genesis chapters 6 to 9. Today, Noah is largely relegated to colourful pre-school storybooks, but Noah's ark is actually about God's saving purposes at a time of climate change. It is a story of sin, judgment and salvation. The story begins with God's sadness and anger at humanity's wickedness, but his initial idea of totally wiping out mankind and all the other animals (Genesis 6:5–7) is soon replaced by a plan combining judgment and rescue. He finds Noah, 'a righteous man . . . who walked with God' (6:9), and instructs him to build an ark which will contain both people and representatives of every living creature. When God's judgment cleanses the earth and kills virtually all life through the flood, God's mercy is shown through those who are included in the ark.

Throughout history, Christians have often seen the ark as a symbol of the salvation that would eventually come through Christ, when God would finally and completely deal with sin and evil on the cross. However, who was included in God's saving purposes? Only eight human beings (Noah, his three sons, and all their wives), but countless other species, each of them precious to God. Genesis 6 – 9 is a highly relevant story for us, because it shows that creation is not simply the backdrop to God's relationship with people. God has saving purposes that include the non-human creation, emphasized in the Noah story in two ways.

Firstly, the reason all these creatures are included on the ark is not that they are valuable to Noah, but simply 'to keep their various kinds alive throughout the earth' (Genesis 7:3). In other words, their value is completely independent of human beings. God didn't include dogs so that Noah could have a pet to keep him company, or cattle so he could enjoy a diet of roast beef. These creatures matter, not because they're useful to us, but because they matter to God.

Secondly, when the flood is over and God sends a rainbow as a sign of his saving covenant, just who is included? It is God's saving promise to Noah, his descendants, and also 'every living creature on earth' (Genesis 9:10). I must have read the story of Noah over a hundred times without noticing this, and yet it is repeated seven times in Genesis 9! In verse 13, God puts it even more bluntly: 'I have set my rainbow in the clouds, and it will be the sign of the covenant between me and the earth.' How dare anybody say that God does not care about the future of this planet? God has a unique and precious covenant with the earth itself.

All of this makes sense when we see who Jesus really is in relation to creation. He is not only Creator and Sustainer, but also the Saviour of the earth. Colossians 1:19–20 says:

> For God was pleased to have all his fullness dwell in him, and through him to reconcile to himself all things, whether things on earth or things in heaven, by making peace through his blood, shed on the cross.

When Jesus died on the cross, he did so to restore all the relationships that had been broken by the fall. Through his Son, God bore the full cost of the evil and selfishness that have been unleashed throughout history and across all creation. The Creator became the crucified, in order to bring the whole creation, made and sustained in love, back into restored relationship with himself.

Remember the triangle of relationships between God, people and creation? Just as all these relationships that God created so good have been shattered, so through the work of Jesus each of these relationships can be restored. In Colossians, Paul is clear that it is 'all things' that are reconciled, or brought back into relationship, with God through the cross. The Greek word for 'all things' (*panta*) is completely inclusive. It includes things both in heaven and also here on earth, the same earth that was cursed in Genesis 3 after Adam's sin and has been groaning in agony ever since.

Romans 8:19–22 compares the 'groaning' within creation to the labour pains of a pregnant woman. Creation is in agony, writhing about as a result of humanity's disobedience. Yet, like a woman's labour, it is a hopeful suffering, looking forward to the new life that is about to be delivered. In the midst of environmental disaster and despair, there is still hope for the natural world. Paul describes it as creation's 'eager expectation' (verse 19), infused with the 'hope that the creation itself will be liberated from its bondage to decay and brought into the glorious freedom of the children of God' (verses 20–21). Just as Jesus' death and resurrection offer hope for human beings and we can become God's children, so there is the sure hope that the natural environment – the whole created order – can be set free.

All of this is possible only because of the cross and resurrection of Jesus. In his book *Jesus and the Earth*, Bishop James Jones has pointed out how the Gospel writers record creation's reaction to Jesus' suffering. When he died, there was an eclipse, as the skies remained dark throughout his crucifixion. At the moment of death there was an earthquake, as tombs flew open and the great curtain in the Jewish temple was torn from top to bottom. These signs show us how the earth itself reacted to the death of the One in whom all things hold together. Then, at Jesus' resurrection, the great stone was rolled away by another earthquake, as creation responded in celebration to the beginnings of a new creation in Jesus' risen body.

Ultimately it is that resurrection body of Jesus that is our guarantee of hope for the material universe. The risen Jesus was neither a ghost nor an illusion. He was, and is, physically alive again. There was no dead body in the empty tomb because it had been transformed into a new resurrection body. The risen Christ is the guarantee that those who trust in him will be raised from death, and the whole created order can also be transformed and renewed.

Ultimately it is that resurrection body of Jesus that is our guarantee of hope for the material universe.

In Jesus' resurrection body there was both continuity and change, and this gives us clues about the future of creation. The risen Jesus still bore the marks of the spear and nails (he showed them to Thomas), so this was no off-the-peg brand-new body clothing a spiritual Jesus. The new body was physical enough to walk a dusty road to Emmaus, to cook and eat fish beside Lake Galilee. Yet there was something different about the risen Christ. He seemed to appear and disappear almost at will, on one occasion appearing in a locked room. So there was both continuity and discontinuity. These two themes will be kept together as we turn next to the final act of the biblical drama: the implications of Jesus' saving work for the future of the earth.

Questions

1. In what ways has this chapter affected your understanding of who Jesus is? How could this influence your devotional life and your lifestyle?
2. Why does the story of Noah (Genesis 6 – 9) seem so relevant today? List some ways in which it helps us understand God's purposes and our response in today's environmental situation.
3. The cross and resurrection of Jesus are the centre of Christian faith. How do they give us hope, for ourselves, and for the planet?

5 The new creation: On earth as in heaven

The resurrection of Jesus marks the final stage of his life on earth. At the same time, it signals the beginning of a new and final act of the great biblical drama. With the risen Christ, the new creation has begun. His transformed and glorified body gives us hope for a physical, material future, both for ourselves and for the whole created order. So it is that we enter act 5, the part of the story over which Christians have argued and differed more than any other.

I want to make clear at the start of this chapter that this is also the area where I have had to rethink my own views most thoroughly. I grew up believing that 'heaven' is somewhere totally disconnected from the here and now, a place beyond space and time, which bears no relation to this world. I was taught that it is a place where those who believe in Christ will go when they die (or when Jesus returns if that comes first), and where we will stay for eternity. As for this earth, I believed that this world would be completely destroyed when Jesus returned at the end of time to judge sin and evil. I was convinced that this was what the Bible taught.

Today I think rather differently, and it is important to state why. I have not changed my views because I've stopped believing the Bible.

Rather, the opposite is true: it is precisely as I have looked at the whole Bible (rather than just one or two slightly tricky passages) that I have come to see that God has far more positive purposes for the earth than I had ever imagined. Understanding the Bible's teaching in this area is a bit like looking at a huge map. Examining small details of the scenery (individual verses) is only helpful after we have got a general overview to gain perspective and some major signposts to guide us. To start with the details may lead us into getting completely lost and confused. More worryingly, if we do not take our directions from the whole of the biblical map, we will have no means of telling whether or not we are starting and finishing in the right place.

The biblical map gives us some very clear co-ordinates to guide us regarding God's purposes for creation. It begins with a God who makes everything good. It continues, despite the disastrous sidetrack caused by sin and the fall, with a God who knows creation intimately, sustains it by his loving power, and sees it as revealing his character. All this points to God's continuing commitment to creation. Although God brings judgment on a world that is spoiled and damaged by sin, his judgment is always discriminating. He never condemns the innocent but only punishes the guilty, and it is clear that human beings, rather than the earth and its creatures, are the guilty parties. At the same time as bringing judgment, God also always offers a means of escape or a method of salvation. At the time of Noah, God showed these principles of judgment and rescue in destroying most of creation, yet rescuing representatives of both humanity and every other living creature. He demonstrated that his saving intentions included all creation, both by the passenger list for the ark and, more remarkably, in who was included in the covenant promise. God's saving purposes included the earth and all its creatures.

Later, as his chosen people settled into their promised land, he provided a framework of laws which included a significant ecological dimension. The people were to care for the land and it would care for them: animals, plants and the land itself deserved respect and sabbath rest. These were not the laws of a God who planned to destroy the earth, but of one who is committed to it.

In sending Jesus, God blessed the material universe by becoming part of it. Jesus lived out a perfect human life, showing his lordship over creation, but also living in harmony with it. Finally, when Jesus died and rose, we have seen how the New Testament is clear that he died to restore all of the relationships broken by the fall. Because of Jesus, creation too is to be set free from its bondage to decay.

The map is clear and it points in one direction only: God's constant commitment to, and saving love towards, the whole created order. There is no place for God utterly destroying the earth in order to replace it with a brand new heaven. Let's change the picture back to the one we have followed in this book, that of the Bible as a play in five acts, with the first four telling of a relationship of love betrayed by humanity and of the terrible impacts on both people and creation. They show God continually seeking ways to regain that love and restore those broken relationships, finally sending his Son to die and rise again. All the way through, the three strands of God, people, and the earth with its creatures have been woven together, building towards a powerful conclusion of restored harmony. It would be unthinkable that at this stage God would suddenly decide to abandon his creation and focus only on human beings. What could be the source of such an idea? Perhaps, rather than originating in the biblical story, it came from the same human beings who have damaged and destroyed creation so terribly. The same blindness that has enabled us to destroy so much of the goodness of God's earth has also led us to justify our behaviour by wrongly interpreting the Bible.

These are strong words, but before we look in detail at what the Bible really says about the subject, it is worth mentioning that Christian history supports this interpretation. From the earliest times, right up until the nineteenth century, the majority of Christians believed that God's plans for the earth were more about continuity than discontinuity, more about a hopeful future than destruction.

I have written some of this book whilst staying on the small island of Bardsey, off the coast of North Wales. Bardsey is famous for its wildlife and for its Celtic Christian heritage, which probably goes back 1,500 years. In a short book about Bardsey's history, A. M. Allchin writes, 'In the Celtic Church there was a very strong sense that ... from the very beginning God had destined his whole creation for resurrection.'[1] The Celtic Christians could not imagine that God would simply destroy the beauty and majesty of a creation that speaks so loudly of his creating and sustaining love.

Rather, the idea that God might remove this earth totally seems to have developed at roughly the same time as the growth of large cities, where human busyness shuts out God's voice in creation, and as the economic ambitions of Western nations led to an increasingly rapid destruction of the earth's resources. Are these two linked? I leave the reader to judge. Meanwhile, it is time to look at the Bible's descriptions of the earth's final destiny.

A vision of harmony in creation

The Old Testament gives us our first indicators of what we can look forward to for creation. The prophets were not all about doom and gloom; they also foresaw a time beyond God's judgment, a time of social and ecological harmony. Hosea 2:16–23 speaks of a day when wars will end (bow and sword abolished), relationships will be equal (wives calling their spouse 'husband', not 'master'), and there will be a new covenant between humans, birds and animals. The broken relationships between people, creation and God will be restored: the prophet speaks of God responding to the skies, which respond to the earth, which responds with grain, wine and oil, and which in turn respond to the people, before finally the circle (or triangle!) is completed as God and the people commit themselves to one other again.

A similar vision appears in Isaiah 11 and 65. In 11:6–9, we read of a time when 'the earth will be full of the knowledge of the LORD as the waters cover the sea'. Just as God's presence filled the garden of Eden, so we are promised that the whole earth will once again be full to brimming over with the reality of knowing God. Like Hosea, Isaiah foresees ecological harmony within all the orders of creation: wolves and lambs, leopards and goats, calves and lions and little children; cattle and bears, lions and oxen, infants and poisonous snakes. None of these will harm one another any more. In Isaiah 65:17–25, similar words are repeated, with further detail about what this will mean for human beings – long life, fruitful work, abundant harvests – all seen specifically as part of God creating 'new heavens and a new earth' (65:17). Yet again the three key relationships, between people, God and the rest of creation, are each seen as being restored.

What are we to make of these extraordinary passages? Some people have tried to spiritualize them completely, saying that they simply point to what heaven will be like. However, one thing is absolutely clear: whether these visions are fully literal or not, they are very tangible, very physical, and very earthy. God's new heavens and new earth are not going to be some weird disembodied state where we commune with God in a heavenly vacuum. These passages are speaking of a material new creation, which has many, many similarities to our current world, but without pain and suffering, conflict and injustice. There are still plenty of unanswered questions. It is hard to understand how creatures such as lions, leopards, bears and wolves will one day cease to be meat-eating predators, when their teeth and digestion seem designed that way. However, we must not underestimate God. The one who created all things good in the first place is surely able to re-create. If I, a sinful fallible man, will one day be made perfect, it will require a transformation no less radical than a lion or a leopard taking up vegetarianism!

The kingdom of God

There are many biblical terms used to speak of the future destiny of all things. 'Heaven', 'the new creation', 'the kingdom of God', and 'the new heavens and new earth' are amongst the most common. It would not be quite accurate to say that all of these refer to the same thing, but in all of these there is one important and perhaps surprising linking theme: the future has already begun! This was hinted at right at the start of this chapter in saying that Jesus' resurrection marked the start of the final act of the biblical drama. The phrase that Jesus himself used most frequently was 'the kingdom of God', or in Matthew's Gospel, 'the kingdom of heaven'. The two mean the same thing: Matthew uses his phrase for exactly the same occasions where the other Gospel writers use 'kingdom of God'. It is very clear if you read through all the times when Jesus uses these words that this kingdom begins with him and also that it is not yet here in fullness. So, Jesus can proclaim the kingdom's arrival from the start of his ministry (Mark 1:15) and speak of the kingdom as 'here for sure' (Luke 11:20, *The Message*), while at the same time speaking of the future feast in the yet-to-come kingdom of God (for example, Luke 13:29; 14:15), and, at the last supper, about these events finding their full meaning in the future kingdom of God (Luke 22:16–18).

God's kingdom is both 'now' and 'not yet'. The kingdom is about God's rule, his reign in individual hearts and lives ('The kingdom of God is within you', Luke 17:21), but also his transforming, healing rule in every dimension of creation: spiritual, physical, mental, social and environmental. The 'gospel', literally the good news of Jesus Christ, is not only spiritual but is referred to as the good news of the kingdom of God (Mark 1:14–15). Thus forgiving sins, healing the sick, casting out demons, restoring self-worth to the poor and stilling creation's storms are all part of the gospel and signs of the kingdom.

With Jesus, the healing of all the relationships broken at the fall has begun, something his actions powerfully demonstrate. The kingdom of God is now. Yet, until sin and evil are completely removed from creation there will still be broken relationships; decay, destruction and death will remain. The kingdom of God is not yet. One glorious day, however, when Jesus returns as king and judge, his kingdom will be fully established and every broken relationship will be healed. Therefore we can agree with Professor Hans Küng, the Roman Catholic theologian, that ultimately 'God's Kingdom is creation healed'.[2] The day will finally come when 'the kingdom of the world has become the kingdom of our Lord and of his Christ, and he will reign for ever and ever' (Revelation 11:15). What a wonderful day that will be!

New or renewed?

The next important misunderstanding to clarify is over the Bible's use of the word 'new'. In English we only have the one word, but in the Greek New Testament there are two: *neos* and *kainos*. While there is some overlap between their meanings, there is also far more variety than in the English word 'new'. For instance, if I talk about my 'new' car, most people will picture a gleaming brand-new model fresh from the factory. However, if cars had been around in New Testament times, then a 'new' car could just as easily mean one that was restored, repaired and effectively recycled. In other words, the word 'new' (particularly *kainos*) can mean 'renewed' rather than 'replaced'.

This is vital in understanding all the biblical language about 'new creation' and 'new heavens and new earth'. The 'new heavens and new earth' referred to by Peter (2 Peter 3:13) or the book of Revelation (Revelation 21:1) do not necessarily imply that the current universe is thrown on the scrap heap (which is after all a rather twentieth-century concept). Rather, they speak of the renewal of creation. This makes perfect sense when we remember that the New Testament also speaks of Christians as 'new creations in Christ' (2 Corinthians 5:17). Does this mean that if I become a Christian, my old physical body is thrown away like a cast-off snakeskin, and I now grow a different biological set of clothing? Of course not! I am the same flesh and bones and DNA as before, but in God's eyes I have indeed become a new creation, and a process of transformation has begun. At present this is invisible, but one day (when Jesus returns and God's kingdom is fully here) I will become a new person, not brand new, but renewed and restored. Just as God is into recycling broken, spoiled, messed-up people, and making them into new creations in Christ, so this whole damaged and groaning creation will be made new again.

It is worth noting that God speaks of his ultimate plan in terms of 'I am making everything new!' (Revelation 21:5). If God were going to start again from scratch, surely he would speak of making lots of new things, rather than making every (existing) thing new? Like a great sculptor restoring a damaged work of art, God is going to remove all that is corrupted by evil and sin and re-mould all that is good and beautiful and right within creation by making everything new again.

Destroyed or purified?

Among the things that confused me for many years were the Bible passages that appeared to speak about the earth being destroyed. Although I could see the logic of everything this book has argued, I had a problem, because certain passages appeared to state quite bluntly that God was going to destroy the earth anyway, however good or important it might be. Let's look in a bit of detail at Matthew 24:1–35, 2 Peter 3:1–13, and some verses from the book of Revelation for examples.

First, it is important to say a word about the kind of literature these passages (along with biblical books such as Daniel and Revelation) represent. These are known as 'apocalyptic' and are a style of writing about the near and distant future that would have been very familiar in the first century but can easily confuse modern readers. Apocalyptic writing often has more than one layer of meaning and is full of symbolism and references to other (usually Old Testament) literature that is vital for interpreting it. Expecting a modern reader who is not steeped in the rest of the Bible to make sense of these passages is a bit like expecting a child who has just learned to read to do *The Times* cryptic crossword. To put it another way, if somebody from a very sheltered English background sits down to watch a Bollywood movie, they may get very confused! It's not only in a different language, but it is a whole different genre with its own traditions, in addition to all kinds of cultural references to Indian religions, the caste system and a history of previous Bollywood films. So, apocalyptic writings such as Matthew 24, 2 Peter 3 and the book of Revelation are perhaps not the best place to start when trying to understand the future of the earth. It is better to get the big picture first and look, as we have been doing, at the sense of where the whole biblical story is heading before focusing on these passages.

Secondly, one common thread between these passages is that they all contain themes of judgment and salvation, discontinuity and continuity. Most mistakes in interpreting them come when people only emphasize one of these and neglect the other. If judgment alone is emphasized, then people tend to think the earth will be completely destroyed and ignore all the more positive references, but if salvation alone is emphasized people may have an unrealistic optimism and believe that creation is slowly evolving towards perfection. Neither of these reflects the biblical position, which holds together God's anger and mercy, destruction and healing, change and transformation.

For example, in Matthew 24 we read Jesus' most complete statement about 'the end of the age' in a passage that refers both to the destruction of the Jerusalem Temple in AD 70 and also to the end of time. There are references to wars and rumours of wars, famines and earthquakes (verses 6–7). These are destructive events, but are seen as 'birth-pains' in a parallel to creation's labour pains in

Romans 8, suggesting that something good will eventually result. The chapter goes on to speak of persecution, great distress and then signs in creation, including the sun and moon darkened, stars falling and the shaking of heavenly bodies, as the Son of Man (Jesus) appears (verses 29–30). In typical apocalyptic manner, these references to the sun, moon and stars include quotations from elsewhere, in this case Isaiah 13:10–13 and Isaiah 34. If we look at these passages we find that they are all about the cleansing fire of purifying judgment, rather than the destructive furnace of blind anger. In Isaiah 34, the result of God's enemies being destroyed is not a blank canvas on which God can paint a new earth, but a landscape that still includes owls and ravens in the ruins from which sinful people have been removed (Isaiah 34:11). God judges in order to enable a new beginning, not in order to completely wipe out.

This is important to remember when we look at Jesus' words in Matthew 24:35: 'Heaven and earth will pass away, but my words will never pass away' (see also Matthew 5:18 and Luke 16:17). At first sight, this seems to predict the complete disappearance of the current Earth and solar system, but Jesus' hearers would have spotted the reference to the Psalms. In Psalm 119:89–90, we find the only other biblical reference linking the heavens and earth to God's word, and it states: 'Your word, O Lord, is eternal; it stands firm in the heavens. Your faithfulness continues through all generations; you established the earth and it endures.' In other words, the lasting nature of God's word, the fact that it will never pass away, is not being contrasted with a temporary throwaway earth, but compared favourably with the most stable, unchanging, enduring thing in creation, the earth! It is bit like somebody saying, 'I will love you until the Rock of Gibraltar falls into the sea'. This is not a prophecy of the end of the Rock of Gibraltar, but a statement of everlasting love! There may also be a reference in Matthew 24:35 to Psalm 102:25–26, which speaks of the earth and heavens 'wearing out like a garment'. Again, this sounds very alarming at first sight, but we're reading as modern people who throw away old garments, rather than using them to make something new. The great sixteenth-century reformer Calvin said of the earth and heavens in this passage that, 'although they will not be completely destroyed, the change of their nature will consume that which is mortal and perishable, in order that they may be renewed according to Romans 8.'[3] There is cleansing and refining judgment in order to get rid of all traces of sin and evil, but what is left is re-used and re-shaped into God's new creation.

If we move on to 2 Peter 3:13, we are faced with a passage that in my experience is most often used as a 'proof text' by those who see the earth as doomed. In the Authorized Version (which was the main English translation until the mid-twentieth century), verse 7 speaks of the heavens and earth 'being reserved unto fire against the day of judgment'; verse 10 states that 'the heavens shall pass away with a great noise, and the elements shall melt with fervent heat, the earth also and the works that are therein shall be burned up'. Finally, verse 13 talks of a 'new heavens and a new earth, wherein dwelleth righteousness'.

No wonder people get confused! However, once again, the Old Testament background is critical. Mention of 'fire' and 'burned up' would not have taken people to images of exploding planets, but to Malachi 3:2–3 where God's judgment is seen as a refining fire, purifying and cleansing, not destroying but leaving the final result pure and without blemish. What is destroyed is not the earth, but 'ungodly men' (verse 7). Similarly in verse 10, poor and confusing English translations from the original Greek have a lot to answer for. The 'elements' that will melt with heat are not the iron or carbon of the periodic table but the elemental spirits of this world.[4] It is the distorted powers that have turned against God and prevented his righteous and just rule that are to be destroyed, so that God's kingdom rule might be fully established. The other key term in 2 Peter 3:10 is 'burned up', which is much better translated (and this is followed in most modern translations) as 'laid bare' or 'disclosed'. Once the judgment has taken place and the fallen powers have been destroyed, the earth will be revealed again.

In conclusion, the true meaning of 2 Peter 3:10 can be summarized as 'The day of the Lord will come like a thief and the heavens will pass away with a loud noise and the elemental powers will be removed through fire, but the earth and the works upon it will be revealed.' Like a good surgeon, God will remove all the cancerous growth caused by sin in our world so that a healthy new earth can be established. No wonder 2 Peter 3:13 looks forward to the new (renewed!) heavens and earth, 'the home of righteousness'.

There is one final clinching argument from 2 Peter 3 to show that this theme of continuity and restoration after judgment is what the author had in mind. In verses 5 to 7 the coming destruction is compared to the flood at the time of Noah. The writer states that the fire of judgment this present world will face will be just like the floodwaters which 'destroyed' the earth at the time of Noah. Yet of course the earth was not completely and utterly destroyed by the flood: it was cleansed, purified, and finally laid bare, so that a new start – a new world – could begin again. When understood properly, 2 Peter 3 actually gives us great hope for the future of this planet, a future that will include great trauma and destruction but will finally lead to a gloriously renewed and restored creation.

Turning to the book of Revelation, right at the end of the Bible, we have a book that is famously difficult to interpret, so much so that many commentators seem quite confused about the future of the planet. Some places seem to imply continuity, others suggest complete destruction. As we have seen, it is in the nature of apocalyptic writing to hold the two themes of judgment and re-creation in tension. A couple of examples will illustrate this. In Revelation 6:12–14, there is a prediction which includes stars falling to earth, the sky rolling up like a scroll, and every mountain and island being removed from their place. Revelation 20:11 speaks of earth and sky fleeing from God's presence. Both passages use the language of violent cosmic change and upheaval.

This is no cosmetic makeover, but a complete remaking.

For God to transform a creation infected by evil's entry into the universe is not simply a matter of re-arranging the furniture, but of completely clearing and rebuilding the whole house. Any notion that 'things can only get better', as humanity and the earth seamlessly evolve onto a higher and better plane, is shown up by this imagery to be no more than a romantic dream. This is no cosmetic makeover, but a complete remaking. Equally, any idea that the earth will be totally destroyed and discarded is also shown to be false. What we are left with is a future that is both terrifying and hopeful. The future of the earth will contain drastic change – the labour pains will get far worse – the earth as we know it now, damaged and defaced by humanity's sin, will not continue as it is for ever. There are warnings to be heeded so that we are ready for Jesus' sudden and surprising return, and will not be caught in the fire of judgment. Yet all is not lost. There is a vision of calm beyond the storm, of gold emerging from the fire, of a new creation emerging from the ashes of the old.

The transforming vision

The great drama of the Bible begins in Eden with a perfect garden and ends in Revelation 22 with a garden city. There are many deliberate parallels. The tree in Eden from which Adam chose disobedience is replaced by a tree of life which gives fruit every month. Its leaves bring healing to nations made ill by sin. The river of the water of life, foreseen by Ezekiel and fulfilled in Jesus,[5] replaces the four rivers of Eden. The city itself represents, not stressful, polluted, twenty-first-century urban life, but a harmonious blending of God's creation with the very best of human creativity, ingenuity and technology. Most importantly of all, the curse, which infected the triangle of relationships between God, humanity and the earth, will no longer exist, because the death and resurrection of Jesus have defeated it. Because of the separation caused by sin since Eden, God and humanity have not been able to walk together within creation, but now God himself will live in the city and people will see him face to face.

From Adam's disobedience until the coming of Jesus, God's perfect rule in heaven remained separate from this earth. With Jesus, the kingdom of God began to be established again, and signs of that kingdom continue to be found. However, when the new creation begins and the New Jerusalem descends from heaven, there will be no more separation between heaven and earth. We don't need to 'go to heaven', because God's home is now with mankind. Tom Wright puts it like this: 'The Christian hope is not simply for "going to heaven when we die", but for new heavens and new earth, integrated together'.[6] This is the transforming vision, that one day, when Christ has returned, 'the kingdom of the world [will have] become the kingdom of our Lord and of his Christ, and he will reign for ever and ever' (Revelation 11:15).

It is impossible for us to imagine exactly what the new creation will be like. It seems that the Bible is deliberately vague on the detail, and so we should be too – after all, we must allow God to have some surprises in store! What we can be certain of is that the surprises will all be good ones. The very best of this world – its most inspiring sunsets, most breathtaking landscapes, most colourful wildlife, most intimate friendships and most exhilarating moments – can only give us a tiny foresight of what is to come. When Jesus is back in his rightful place as Lord of all creation, our joys will all be complete, and yet we will only be discovering the start of what joy really means.

Questions

1. How has this chapter challenged your understanding of 'the end times' and the future of the earth? List any unanswered questions you still have, and (using the further resources section or in discussion with others) try and resolve these ... whilst recognizing that the Bible does not give neat answers to all our questions!

2. If the kingdom of God is both 'now' and 'not yet', what signs of God's kingdom can you trace in your experience, and in the world around?

3. Read some of the biblical passages that describe the new creation (such as Hosea 2:16–23; Isaiah 11:6–9; 65:17–25), and then try to write your own description of a world where harmony exists between God, people and the rest of creation.

6 Living it out: Discipleship as if creation matters

So far we've had an overview of the great drama of God's relationship with his creation. We have seen that caring for creation is not insidious, irrelevant or incidental to Christian faith, but intrinsic to the good news of the gospel. If we are convinced of this truth, it cannot remain in our heads but must start to impact on our lives. The rest of this book explores how we are to live this out in our discipleship, our worship, our lifestyles and our mission.

'Discipleship' is a favourite topic of preachers and ministers. Christianity is not simply a matter of accepting certain beliefs, but of being a follower (the literal meaning of 'disciple') of Jesus Christ. Traditionally, sermons or books on discipleship have focused on certain things, with prayer, Bible-reading and church attendance being the favourites. These are all vitally important: prayer is talking and listening to Jesus; studying the Bible is grappling with his teaching; and church is (at its best!) spending time with his family. Sometimes teaching on discipleship goes on to focus on moral and ethical attitudes that Christians should have. However, it is rare for discipleship training to include much on our relationship with nature, the land and our fellow-creatures. Yet, as we've seen, these are central biblical themes and we cannot be true disciples of Jesus without our relationship with our environment being affected.

Imaging God

Human beings play a vital role in the drama of God's relationship with creation. As we saw back in the first act, we are both part of creation and called apart within creation. We are the cause of the fall and responsible for the groaning of creation and the breakdown of relationships within the whole created order. However, it was also through a human being, Jesus, that God's saving plans were brought about, giving us hope both for ourselves and for the whole creation.

You may remember that in Genesis chapter 1, human beings are described as being made in the image of God (verse 26). This chapter aims to explore just how we can discover our true human nature as God's image by being disciples of Jesus Christ in our relationship with creation.

The phrase 'image of God' has fascinated philosophers and biblical scholars down the centuries. Thousands of pages have been filled trying to work out whether this refers to our physical appearance, self-awareness or conscience, or the mental or spiritual capacities that separate us from animals. All of these have their merits as philosophical ideas, but none are rooted in Genesis 1, where the phrase first appears.

Here is Genesis 1:26–28 from *The Message*:

God spoke: 'Let us make human beings in our image, make them
     reflecting our nature
So they can be responsible for the fish in the sea,
     the birds in the air, the cattle,
And, yes, Earth itself,
     and every animal that moves on the face of Earth.'
God created human beings;
     he created them godlike,
Reflecting God's nature.
     He created them male and female.

God blessed them:
  'Prosper! Reproduce! Fill Earth! Take charge!
Be responsible for fish in the sea and birds in the air,
  for every living thing that moves on the face of Earth.'

It is clear from this crucial passage that being made in God's image is linked to how we relate to the earth and our fellow creatures. Effectively God says, 'I make you in my image so you can be responsible for the earth and its creatures.' Reflecting God in caring for creation is fundamental to who we are as human beings. It is our first great commission and our clear job description.

As we look at the state of the planet today, with an uncertain future due to climate change, over-exploited resources, and species rapidly heading towards extinction, we can see how far we have fallen from the job entrusted to us in Genesis 1. As human beings we have dismally failed to reflect God's image in caring for creation. Not only does this have terrible results for the planet; it also damages us as human beings. When we fail to care for the earth in a godly way, we fail to reflect God's image. If we neglect the planet, we become less truly human and God's image in us begins to fade away. There is a prayer of confession in the Anglican communion service which says, 'We have wounded your love and marred your image in us.'[1] As the human species, our destructive relationships with God, one another, and the rest of creation have badly blurred and damaged God's image. We no longer reflect it, and desperately need help in modelling what it means to be truly the image of God in human skin.

Jesus and the image of God

Jesus is our perfect model in terms of being God's image, and if we become his disciples we can learn from him. Paul describes him as 'the image of the invisible God, the firstborn over all creation' (Colossians 1:15). The word used here for 'image' is *eikon* in the original Greek, which means a visible representation of some-thing invisible. An icon is a window into another world, a glimpse of a greater and truer reality. Eastern Orthodox Christians have always seen their religious icons, beautiful pictures of Jesus or of Christian saints, not simply as portraits but as objects with the potential to transform us spiritually by drawing us into their world.

Jesus is the perfect image of God in his relationship with creation. Firstly he showed God's commitment to a material physical world by becoming part of it. Becoming a human being (the incarnation) is God's strongest possible 'yes!' to his creation. Secondly, no other human being has lived as Jesus lived, in perfect unbroken relationship with God, other people and creation. The life of Jesus gives us a perfect example of how to 'image' or reflect God in our relationship with creation. He is the perfect 'icon', showing us God and godly relationships with all that is around him. Like all great icons, Jesus draws us into his world, inviting us to follow him in his attitudes and actions.

Another area where we can easily miss Jesus' real significance is in his frequent reference to himself as 'the Son of Man'. This isn't just about his humanity or a nod to some slightly obscure Old Testament passages. It is most obviously a reference back to Adam, the first man. 'Son of Man' is literally 'Son of Adam' . . . and Adam (as we saw in act 1) means 'created from dust'. Jesus is talking about himself as 'the son of the one hewn from the earth'.[2] Similarly, when Paul talks about Jesus as 'the second Adam' in Romans, he is referring to the second 'son of the soil'.

Whilst the first human failed to reflect God in his relationship with the earth, this second Adam, 'son of the earthling', is the perfect image of the invisible God. He is therefore our perfect model in terms of creation care. He had power over creation: turning water into wine, multiplying bread and fish, and mending broken bodies and minds. Yet he also cared for creation. Jesus said that we are worth more than many sparrows, yet he also said that not a single sparrow falls to the ground without God's knowledge (Matthew 10:29).

Most of all, Jesus demonstrated a form of leadership that perfectly illustrates how we are to image God in our relationship with creation. In Genesis 1, the older Bible translations referred to humanity as 'having dominion' over the earth and its creatures. 'Dominion' sounds very close to domination, and has sometimes been used both by Christians to excuse exploiting the earth and by critics to accuse Christians of using the earth's resources selfishly and wastefully. However, Jesus shows us that being God's image is not about power and privilege, but about self-giving service.

The real meaning of 'dominion' in Genesis 1 is 'lordship', the just and gentle rule of a godly leader or parent. For Christians, Jesus models what lordship and kingship are all about. He was the servant king who gave up all the glories of heavenly living to be born in poverty and grow up in a humble home. In modelling leadership to his disciples, Jesus wrapped himself in a towel and washed their feet, illustrating his own words, 'The Son of Man [literally, the Son of Adam – the earthling] did not come to be served, but to serve' (Matthew 20:28), and 'Whoever wants to become great among you must be your servant' (Matthew 20:26). Jesus was showing us how to image God. If Jesus is God's image to us, and we are called to be God's image in our treatment of the rest of creation, then we should imitate his servant attitude.

Next, as we think about how we can be Jesus' disciples in our relationship with creation, we should note how he celebrated the goodness and provision of creation. He was not a barefooted, sackcloth-wearing monk. He made hundreds of gallons of excellent wine at a wedding. He shocked the religious authorities by allowing his followers to pick and enjoy food even on the day of rest, the sabbath. He described God's kingdom in terms of food and drink and feasting. Jesus taught us how to use and enjoy creation without abusing and exploiting it: he was the ultimate ethical consumer! He lived very simply. Of course, it was the culture of his time to live off the land. Yet Jesus also enjoyed the bounty of creation, and knew how to party and celebrate.

Today, in a very different world where population growth and obscene over-consumption mean that a few have far too much and many struggle on less than enough, how do we understand Jesus' example? I believe we should live as close to the land as we can, becoming more aware of God's provision through the seasons and our dependence on rain and sunshine. We should consciously thank God for our food, enjoying it and celebrating it. At the same time, we should be good stewards, not wasting or over-consuming, lest Jesus' stinging attacks on hypocrites whose lifestyles deny their beliefs come back to haunt us.

Understanding the image

So, if we are to follow Jesus and reflect God in caring for creation, what does this mean in practice? The Bible uses a number of helpful pictures, the first one coming right after creation, when God sends Adam into the Garden of Eden to 'work it and take care of it' (Genesis 2:15). The words here can well be translated as 'to serve and to preserve'. In other words, we are to be **gardeners** and **caretakers** within God's creation. A gardener works or serves the garden by planting seeds, training plants, watering and pruning, and harvesting good produce. In Genesis 1:28 humans are instructed to 'fill the earth and subdue it'. Subduing is something at which humanity has excelled. Gardening does involve subduing the earth, fighting back weeds, managing and developing creation. However, a gardener subdues in order to bring the best out of nature, not to destroy it and make it infertile. A gardener manages and controls the garden in a sensitive way, neither leaving it alone as a wilderness, nor pouring concrete over every part of it.

Caretakers preserve the premises for which they are responsible. They protect from harm and they keep watch. As we image God, we are to serve and preserve the whole earth and all its creatures. We are permitted to enjoy the fruits of creation and develop God's world for productive use. Yet we are also to be restrained in how we do this, preserving an ecological balance between species, ensuring that our use of resources does not threaten particular species or ecosystems.

Another very biblical image is that of being **tenants** and **stewards**. We've already seen how the Earth belongs to God, not us, and that our place is as leaseholders or tenants, living on God's property and answerable to him. In the Gospels, Jesus tells several stories about being good stewards or managers. Often these parables are seen as being only about money, but actually they are about managing all that God has made and entrusted to our care. A steward is somebody who looks after property on behalf of someone else.

In Matthew 25, Jesus tells a story about a landowner giving three stewards different amounts to look after, according to their abilities. When the landowner returns, he is pleased with those who have made a healthy profit, and furious with the one who has simply buried his share. How are we to understand this in terms of our stewardship of creation? A good steward, according to Jesus, not only protects what he or she is given in trust, but develops it for the good of all. We are not to conserve creation like a great theme park with wildlife reserves and refuges where ever-dwindling wild species hang on by their fingertips. We are to encourage creation positively, to develop it creatively and carefully, not just for our own selfish human good but for the good of all species. Used rightly, and in the service of a sustainable vision, science, technology and business can all be expressions of this good stewardship, enabling creation to reach its full potential. However, when science, technology or business is used selfishly or as an expression of human mastery, it can be amongst the most destructive forces we can unleash.

Like all managers or stewards, we may be called by the owner to give an account of how we have used or abused what has been entrusted to our care. One particular passage that has always challenged me deeply is where Jesus says: 'From everyone who has been given much, much will be demanded; and from the one who has been entrusted with much, much more will be asked' (Luke 12:48b). Those of us who live in Western societies have indeed been entrusted with much in terms of material wealth and natural resources, and we are answerable not just to the world's poor and to our fellow-creatures but to God for how we use and develop them.

In the Old Testament, the three groups who were set apart to have a special role standing between people and God were **prophets**, **priests** and **kings**. Prophets proclaimed God's word, often to a world that didn't want to hear. Priests offered the worship of the people to God. Kings represented God's authority and rule over the people. In the New Testament, all three roles are fulfilled in Jesus Christ: the prophet who told us about God, the great high priest who offered in his death a sacrificial offering that need never be repeated, and the king who now reigns over all. Furthermore, Jesus' followers, imitating him and walking in his footsteps, are also variously described as prophets, priests and kings. Although the Bible does not explicitly use this language to talk of our human relationship with nature, it is an analogy that works extremely well.

We are to have a *prophetic* role in terms of speaking out about the true damage that our lifestyles and behaviour are causing to the planet. Today, an increasing number of Christians are being called to this. In the work of A Rocha[3] around the world, we are finding Christians who are being challenged by Jesus to give up lucrative or promising careers, downsize their lifestyles, speak out against complacency (amongst churches and beyond), and challenge inappropriate and damaging development. Recently I received a letter from a young Christian who had attended a conference where I'd spoken. He was inspired by the biblical message on living sustainably and justly, and went home to found a campaigning website. Just like Jeremiah, Ezekiel and the other prophets telling uncomfortable truths to a people who thought they were doing just fine and didn't need God's help, so we can speak out about the effects of our unsustainable lifestyles on the planet, and on poor people around the world.

We have a *priestly* role, because we can express creation's unspoken worship of God through language and music and creativity. (We will explore this further in chapter seven.) In a real sense, all human creativity – art, drama, sculpture, music, dance – is at best a reflection of God's creativity as expressed in nature. It can be a sacrifice of worship offered back to God which sums up creation's worship and leads us into God's presence.

Finally, we also have a *kingly* role as God's representatives, ruling the world compassionately on God's behalf. As we've seen, this is not about lording it over the earth, but about caring for it responsibly. Psalm 8 sees human beings as both incredibly insignificant and yet set apart by God to rule over creation. Like all good leaders, we must retain a sense of complete humility and unworthiness. We rule over creation, yet are part of it ourselves. We are looking after our own home and that of millions of other species. It is both an amazing privilege and an enormous responsibility. Most importantly, we cannot rule over this world on our own but only in dependence on our Lord, the Creator, Sustainer, Saviour God.

Discipleship today

In conclusion, a biblical understanding of discipleship must include following Jesus in caring for creation. He is our perfect model in his own relationship with creation; enjoying its fruitfulness, developing it creatively (as a carpenter), and living respectfully and gently within it. As we've seen, there are helpful models to understand our role as Jesus' disciples within the world: gardeners and caretakers, tenants and stewards, prophets, priests and kings. None of these is a passive role. All of them require action and involvement, not just in words but in changes to the way we live our lives.

However, the exciting thing that I am discovering increasingly is this: when we step out and seek to image God in creation care, we find that Jesus has already gone ahead of us. He is already at work in his world, not just in the background sustaining the forces of the universe and the changes of the seasons, but preparing the way for his people to help restore his image. If creation is groaning, it is not only in agony but also in longing for God's sons and daughters to rediscover their true place within it. 'The creation waits in eager expectation for the sons of God to be revealed,' writes Paul in Romans 8:19. Jesus, as the first-fruits of a restored, healed creation, is not only our inspiration as we seek to be God's image. He is also preparing good works for us to do in restoring the damaged fabric of relationships within creation. Being a disciple of Jesus is to include serving him through caring for his world, helping restore the relationships broken by sin, and recovering humanity's first great commission.

Questions

1. Using Genesis chapters 1 and 2 and the example of Jesus, write a job description for humanity as 'the image of God'.

2. Our role as disciples of Jesus is described in this chapter as being like gardeners and caretakers, tenants and stewards, prophets, priests and kings. Which of these do you find most helpful or unhelpful, and why?

3. How can we use the gifts of science, technology and business in a way that sustains and stewards creation, rather than exploiting or destroying it?

7 Living it out: Worship as if creation matters

Worship is the purpose of being. It is about the whole of our lives, not just the 'religious' bits such as prayer, singing or Bible study. We can worship God while driving the car, shopping in the supermarket and talking with friends. It is partly about being tuned in to God in the midst of all we're doing, silently dedicating our activities and encounters to God.

It is also about what those activities entail. A person might be worshipping God in their hearts while at work, but if that work includes exploiting the poor, cheating the taxman or polluting the earth, then it ceases to be worship that pleases God. If I walk around the supermarket listening to worship songs on my MP3 player whilst buying meat from animals that have been treated cruelly simply in order to make meat cheaper, then it ceases to be worship of our Creator Sustainer Saviour God.

The following two chapters examine the Bible's wonderfully integrated understanding of worship. Before examining practical issues of lifestyle (see chapter eight), let us start by looking at the context of our relationships with God, one another and the planet.

We will use an outline of the word 'worship' as an exploration of how we can worship God as if creation matters:

W Wonder at God's creation
O Openness to God speaking through creation
R Rootedness in the place where God puts us
S Sabbath rest and re-creation
H Hands-on involvement with creation
I Integration of all our relationships
P Prayer for God's kingdom

Wonder at God's creation

We live in a world starved of wonder. However, just watch an infant class released into a garden or a field, and you will see them bubble over with joy and amazement at what they find. Young children have an inbuilt sense of wonder, but sadly as we grow older it is squeezed out of us as we learn to be logical, cynical and sceptical. We forget how to wonder. We learn the science, and how a rainbow is simply light refracted through damp air, and lose the childlike joy of seeing one magically appear in the clouds. We rush along the road in our busyness, forgetting to stop and notice the puddles, the leaves, the faces we pass. Yet it doesn't have to be like that. The adults I know who still have a sense of wonder are generally the most interesting and profound people I know.

We learn the science, and how a rainbow is simply light refracted through damp air, and lose the childlike joy of seeing one magically appear in the clouds.

Wonder is the stepping stone to worship. It is a short step from reflecting on how beautiful or amazing something is to saying 'thank you' for it. And of course, whether we know it or not, it is the Creator whom we are thanking. God's creation is the setting where we can rediscover our sense of wonder and is God's chosen place for us to start our worship.

In order to rediscover our sense of wonder and transform our worship we may need to change our relationship with God. I find it very strange that it is seen as normal for Christians to sit indoors with their eyes closed in order to worship. Our personalities all vary and some people may find it most helpful to concentrate on God while sitting silently in a room. Sadly, though, it has often been seen as the only 'right' way to nurture a relationship with God. For many people it is liberating and enriching to seek God's presence and pray outdoors while walking, jogging, cycling or sitting quietly with eyes and ears open to all around us. This is nothing new, but is in fact biblically much more 'normal' than worshipping indoors.

The Psalms, the worship manual that Jesus used, love to celebrate the wonder of creation. Psalm 100 begins, 'Shout for joy to the LORD, all the earth.' Notice that it's all the 'earth', not just all the people! Psalm 96 also encourages 'all the earth' to sing a new song to God (verse 1), and goes on specifically to include some of the non-human parts of creation:

Let the heavens rejoice, let the earth be glad;
    let the sea resound, and all that is in it;
    let the fields be jubilant, and everything in them.
Then all the trees of the forest will sing for joy;
    they will sing before the LORD, for he comes,
    he comes to judge the earth.
(Psalm 96:11–13)

There are many, many more passages that speak in similar language. Creation praises God by itself, and there is no better stepping stone for us to worship God than to immerse ourselves in creation's praise. The whole of creation is a massive complex orchestra and each of God's creatures is an instrument, worshipping God simply through living their lives as God intended. Every part of creation worships God when it fulfils the role with which God entrusted it. Psalm 148 lists how great sea creatures, fruit trees and cedars, wild and domestic animals, small creatures and birds, kings and rulers, men and women, old and young, all worship God. Even those things that have no breath or voice praise God. Psalm 98 talks of the whole earth bursting into jubilant song with music:

seas 'resounding', rivers 'clapping their hands' and mountains 'singing together for joy'. In the New Testament, Jesus said that if we were to stop worshipping God, even the stones would cry out to him (Luke 19:40).

As human beings we are both part of this orchestra, instruments designed to play in harmony with others and also called to a special role as God's image-bearers. Our role is rather like that of a conductor, enabling each instrument of the orchestra to play its part in harmony. Where one species is dominating unhealthily, we can tone it down; where another needs protective nourishing to develop its place, as conductors we are to nurture it. We are to draw out and bring forth all that is best from creation. The German theologian Jürgen Moltmann writes: 'in the praise of creation the human being sings the cosmic liturgy, and through him the cosmos sings before its Creator the eternal song of creation'.[1] To rephrase this, we can give voice to the worship bubbling up all around us and offer that worship to God.

Worshipping God through reawakening our sense of wonder in creation is not a replacement for worshipping through Bible study. The first six verses of Psalm 19 talk about the natural world and how it declares God's glory, before going on in the next five verses to talk about how important and helpful the Bible is. We need both: God's book of works and his book of words. God's works, creation, stir up our feelings of wonder and praise, and God's word, the Bible, makes sense of those feelings.

I have found on many occasions that when my relationship with God is dry and hard going, it is a sense of God's wonder in creation that draws me close again. Living in a very urban, noisy, crowded part of London, I find there are still plenty of places where God's world brings me into a sense of wondering worship. I look up, see the ever-changing clouds and think of God's sustaining power in sending rain and sunshine to water the earth and give growth to living creatures. I feel the grass under my feet and see hard-working ants and other tiny creatures, and remember that I too am part of this amazingly complex inter-related world. If I sit too long I may be reminded of Proverbs 6:6 where lazy people who won't try to work for a living are contrasted with busy little ants! I constantly find that as I bring my praise and prayers to God, the natural world inspires, challenges and helps me, reminding me of God's power, creativity, care and attention to detail. If I have worries or burdens, they are often put into perspective as I find my place in creation.

For many Christians down the centuries, there have often been special places that have a deep sense of God's presence in the created world. The Celtic Christians who spread the gospel throughout northern Britain and Ireland often based themselves on islands of great beauty where God's 'invisible qualities – his eternal power and divine nature – have been clearly seen, being understood from what has been made' (Romans 1:20). Places like Iona in the Scottish Hebrides, Lindisfarne, the 'Holy Island' of Northumbria, and Bardsey off north Wales were all seen as 'thin places': prayer-soaked sites where the veil that separates everyday life from the reality of God's kingdom seemed to have worn thin, revealing a deep sense of God's presence. We should not dismiss as superstitious those people who visit such places hoping for an encounter with God. They are in a long Christian tradition fulfilling the biblical pattern of encounter with God through experiences of natural wonder. Of course there's the danger that people develop a superstitious relationship with such places, and start to focus too much on nature and lose sight of the Creator God, who is beyond creation. However, that should not prevent us from seeing beautiful places (whether they be historic pilgrimage sites or wonderful mountains, beaches or woodlands) as great evangelists, bringing people into a sense of God's power and nature.

Wherever we live, whether on a beautiful island or in the midst of a busy city, let us cultivate our sense of wonder. Like a sensitive plant, it withers unless nurtured. If we seek out places where we can see God's footprints and hear his voice through creation, not only may we be led into worship, but we will be more likely to care better for the world around us.

Openness to God speaking through creation

In chapter one we saw how creation speaks of God in many ways, supported by both experience and the Bible. God allows creation to speak to us both indirectly and directly. I have personally experienced on a number of occasions how God can speak through creation, never contradicting what he says in the Bible but bringing its message home with a fresh directness. When I've been worried about situations (health, finance, relationships), I've often been comforted and challenged by observing nature. I try to take literally Jesus' challenge not to worry but rather to study the birds and the flowers (Matthew 6:25–33). As a natural worrier, I've learned so much from seeing how God provides for his creation and also how, along with other animals, we still have to seek and work for our food.

In the book of Job we read:

'But ask the animals, and they will teach you,
  or the birds of the air, and they will tell you;
or speak to the earth and it will teach you,
  or let the fish of the sea inform you.
Which of these does not know
  that the hand of the LORD has done this?
In his hand is the life of every creature
  and the breath of all mankind.'
(Job 12:7–10)

Creation can be God's way of speaking to us when we've lost our sense of perspective. This passage, in which Job is replying to his useless friends, is largely about the sense of dependence in the natural world. Life and breath are God's to give or take away. In our arrogance as humans we sometimes act as if we're immortal or untouchable. You could call it the *Titanic* syndrome, believing that something we've created can defeat the powers of nature or even the hand of God. It's been going on since the Tower of Babel and is still happening today. Every time politicians or scientists proclaim that climate change can simply be 'fixed' by better technology, they fall into this trap. The world is far bigger and more complex than the minds of human beings, and God is far bigger and more complex than the universe he has created!

If we recover a real sense of humility we will see ourselves in perspective. Creation can help us to do that. This is why I believe that both the terrible Asian tsunami of 2004 and the destruction of New Orleans by Hurricane Katrina in 2005 were wake-up calls. Both of these terrible events shook humanity's faith in itself to the core, demonstrating that however clever and advanced we are as a species, we remain a tiny part of a very complex creation. Isaiah 40 reminds us that 'all men are like grass, and all their glory is like the flowers of the field' (verse 6), and even the most powerful nations on earth are 'like a drop in a bucket' or 'dust on the scales' (verse 15).

Mankind is not the measure of all things. In the scale of creation, as we stand on top of a mountain and see the view stretched out, or look at the night sky and see stars that are many light years away, we get a true sense of who we are. We know our insignificance and yet we also know our need of God who amazingly chooses us to follow him and become his co-workers in his world.

God also speaks through creation's groaning at the way we have abused it. Earlier, we saw how Romans 8 describes how the whole created universe 'has been groaning as in the pains of childbirth right up to the present time' (verse 22) due to the breakdown of relationships we have caused by rejecting God's ways. Yet it is only one of three 'groanings' referred to in Romans 8. Along with creation, we who have God's Spirit 'groan inwardly' (verse 23), and God himself through his Spirit prays for us 'with groans that words cannot express' (verse 26). God is not immune to the sufferings of his world. By his Spirit, he joins in our pain and the pain of creation. We can only make sense of the tremendous pain and cruelty within today's natural world (much of it caused by ourselves) and come to terms with human suffering when we understand that God too is intimately involved in the suffering and groaning of the world.

Creation can be God's way of speaking to us when we've lost our sense of perspective.

We can see God's involvement in creation's complexity and suffering in the story of Job. For Job everything is stripped away: comforts, possessions, securities, reputation, relationships, health. He receives no comfort from philosophy, theology or logical attempts to explain his plight, but in chapters 38 to 41 God takes him for a guided tour around his creation. He takes Job to look at the heights of the heavens and the depths of the oceans. He shows him massive weather systems and the tiny details of dew and frost. He looks at all kinds of strange creatures. What is God doing? He is putting Job's questioning in context. 'Where were you when I made this? Could you do that?' He is not trying to crush or mock Job. He is simply allowing him to see the bigger picture.

In creation Job sees a God of astonishing power and majesty, but also of artistry and gentleness, and humour. There is an order and purpose within creation even when Job's own life seems meaningless. There is a hand of love on the tiller of the universe, even if there is also terrible suffering. Job is made to regain a sense of perspective on how his sufferings fit in the massive scale of things and forced into silence. The final theme, only hinted at in Job but made clearer in the light of the New Testament, is that God is intimately involved in his whole creation: he rolls out the dawn and unveils the stars. He is not remote from the mystery and from the suffering, but is deeply involved with it. By the end of the book, Job has no easy answers to his questions, but nevertheless he is reconciled to God. He has met God through his creation.

Today we need to listen hard to God speaking to us through creation. Our noisy lives easily shut out the quietness of creation's whisper, whilst God's damaged world groans louder: in the creaking of melting glaciers, the crackling of forest fires and the roar of hurricanes.

My life was turned around when I first heard God speak through the groaning of creation. I was in the act of throwing away my family's rubbish while holidaying on a beautiful island when God spoke. I could easily have missed it, but an inner whisper said, 'How do you think I feel about what you're doing to my world?' The bin bags in my hand, similar to the ones I threw away unthinkingly at home, suddenly looked and felt different. They symbolized my abuse of God's good world: my taking for granted his provision, my idolatry of greed and pleasure, and my selfish wastefulness. God spoke, creation groaned, and worship could never be the same again.

Rootedness in the place where God puts us

We live in a world of endless mobility and rootlessness. The sociologist Zygmunt Baumann talks of people today as 'tourists' and 'vagabonds'.[2] Tourists are those who can afford to travel for pleasure, flying on cheap package tours to 'see the world' and collect brief and immediate sensations, before moving on to something else. The vagabonds are the have-nots: those who keep moving on because life is too hard where they are. They are the millions today who are refugees, travelling from one place to another, seeking an escape from war, persecution, poverty, or environmental problems. According to the United Nations, climate change means that there will be at least an additional 150 million environmental refugees by 2050. In between are all the ordinary people who move from place to place because of housing, or jobs, or to be nearer to (or further from!) family.

The results of this massive global uprooting are disastrous. As our world has become a place of rootlessness, we see the harmful results in broken relationships: ecologically, socially, mentally and spiritually. It is rare to find people who really know the places they live in, or can talk about the wildlife or when the first blackberries used to ripen. Increasingly, urban children have little idea that what they eat grew in the soil rather than on a supermarket shelf. Food can come from thousands of miles away or from just down the road and often we have no idea.

Yet we are created to belong to a people and a place. We were designed to be in relationship, not only with God, but also with a community of people and the natural world. In chapter three we saw how God created us to belong deeply to the places where he puts us. In Jeremiah 29, God speaks to the Israelites in exile in Babylon, urging them to put down roots. Wherever we are, even if it's not a place we have chosen, God is equally challenging us to pray and work for the welfare of the cities, towns or villages where he has put us. Jeremiah 29:4–7 is such an important passage for us today that I will quote it again:

> This is what the LORD Almighty, the God of Israel, says to all those I carried into exile from Jerusalem to Babylon: 'Build houses and settle down; plant gardens and eat what they produce. Marry and have sons and daughters; . . . Also seek the peace and prosperity of the city to which I have carried you into exile. Pray to the LORD for it, because if it prospers, you too will prosper.'

Today God is challenging Christians to be people who put down deep roots into their local communities and their local environments as part of their worship.

To illustrate this let me explain what I've learned about mangroves. Mangroves are specialized tropical plants which grow in the inter-tidal zones where fresh and salt water meet. They have complex root systems to help them survive in areas of strong currents and tides, and healthy mangrove swamps become places of amazing biodiversity. They provide nurseries for fish and turtles, and nesting sites for birds; the muddy silt they hold together is full of shellfish and invertebrates. They also act as natural shock-absorbers. When tropical storms, cyclones and tsunamis threaten a coast, mangroves absorb much of the impact, thus protecting the soil, animals and villages behind them. There is clear evidence that the 2004 Asian tsunami was at its most devastating in areas where mangroves had been destroyed, but had far less impact where healthy mangroves had remained.

Today the storms our world is facing include not only a changing climate (although that is one of the most serious) but other massive social, economic and environmental upheavals. There is a desperate need for people to be like mangroves: deeply committed to the places God has put them. Deeply rooted people hold together the fabric of a community. Deeply rooted people get to know their local ecosystem: the right things to grow and when to grow them, and they notice when the seasons are changing. Deeply rooted people don't move on when things get difficult, but rather draw on those deep roots into God and the human and natural community that supports them. I am not saying that God never wants us to move from where we now live, but that we should not let this be an easy way out or a means of excusing a lack of involvement in the places where we now are. After all, the people of Israel in Jeremiah 29 were only to be in exile in Babylon for about seventy years, but in God's eyes that was long enough for them to get wholeheartedly involved. If and when we do move, we should seek to put down deep roots and really get to know our new home, both the human community and the natural world.

In a rootless world, Christians believe in a God who put us in a garden and told us to work it, who put a people into a land with instructions to care for it, and sent his Son to spend thirty years getting to know a local place and community. It is an essential part of worshipping this God to get to know the place where he has put us. Bloom where you're planted!

Sabbath rest and re-creation

Isn't it strange how we have turned God's precious gift of time into a commodity? We 'spend' time as if there were a meter ticking and adding up the cost of every second, and we may see enjoying creation as a 'waste of time'. Yet it is within creation that we find a true sense of God's time. God has created rhythms and patterns that we neglect at our peril. They are woven into the fabric of life and if we try to ignore them we become frayed at the edges and worn down. God's time consists of sun rising and setting, moon waxing and waning, tides ebbing and flowing, and seasons revolving regularly. Without creation, our sense of time becomes distorted, focusing on the next deadline (a symbolic word, surely contrasting with lifelines?) rather than enjoying the current moment.

Our society has forgotten a vital aspect of God's creation: the need for sabbath. Sabbath is a time of rest, play and recreation (re-creation!) in God's presence. After six days of hard work creating light and dark, sea and sky, land and plants, sun, moon and stars, fish, birds, animals and people, God 'rested from all his work' (Genesis 2:2). Even God wanted a break.

Sabbath rest is a time to meet together to worship God and be with family and friends. But there is also more to it, a whole dimension that has often been missed. Sabbath is a key to understanding the relationship between God, people and the whole creation. It is not only for the benefit of people but also for animals (Exodus 20:10) and for the land itself (Leviticus 25:1–7), as we saw in chapter three. The people of Israel were clearly commanded to respect their fellow-creatures by sharing a time of rest, and to allow the land itself to enjoy a sabbath every seventh year. On numerous occasions when the Old Testament talks about the sabbath, we are reminded, 'It is a sabbath to the LORD.' The sabbath is as much about creation as about people, and it is all about God!

All relationships need 'quality time'. Just as a couple needs to build in space without distractions to enjoy each other's company and work on their relationship, so sabbath is quality time for the three-way relationships between God, people and the natural world. It is about our relationship with God, with creation, and God's relationship with creation. God delights in creation. He enjoys his own handiwork in all its diversity. He has created it to function according to the rhythms of night and day, ebb and flow, summer and winter, fallowness and fruitfulness, work and rest. Sabbath is about making sure we are connected to these patterns that come from the heart of God. It is about God's commitment to the earth, ensuring that 'as long as the earth endures, seedtime and harvest, cold and heat, summer and winter, day and night will never cease' (Genesis 8:22). It is about the importance of rhythm and rest for humanity, animals and the earth itself.

By nature I am an activist. I tend to fill my diary until I am too busy. Yet I have learned that the worship that God wants is not a diary full of events in which I busily serve him. He wants *me*. He wants my unrushed presence, walking in the cool of a garden (as Adam and Eve did), restoring my soul as I sit beside still waters (as David did in Psalm 23), or going away to a quiet place with some friends (as Jesus did with his disciples). God said through Isaiah: 'In repentance and rest is your salvation, in quietness and trust is your strength, but you would have none of it' (Isaiah 30:15).

Sabbath rest is about rediscovering the rhythms that God wants us to have. These rhythms are essential for our spiritual, mental and physical welfare. They are also vital if we are to live at the pace of the planet, living as part of a healthy creation rather than separate.

Hands-on involvement with creation

Some years ago I came across a survey by a major medical research charity showing a direct link between areas with a lack of access to 'green space' and an increase in mental health problems. At the time I was minister of a church in a heavily urban area and the findings did not surprise me. Since then I've noted several surveys on 'happiness', where people consistently identify one of the key factors in their sense of well-being as 'contact with nature'. God is relational and has created us to relate, not only to himself and each other, but to the world around us. In a busy, built-up area where even gardens had become covered in concrete, I could see a mental and spiritual poverty in the lives of people around me caused by a complete lack of relationship with creation.

In the church garden, my two-year-old daughter pointed out a dandelion to a much older child. 'Don't be stupid,' the older girl scorned, 'It's not a lion!' She had no knowledge of even the commonest flowers. On another occasion a woman from our church looked out of our kitchen window and saw a robin on the bird table. She gasped in astonishment, 'But, but it's real!' Even though they are among the commonest of urban birds, she had thought robins were make-believe birds on Christmas cards and had never before noticed one.

All of us suffer if we spend too long away from God's creation. We are part of creation, and our first God-given tasks were to look after the garden and name the animals. I have come to understand that it is vital for every human being to have some hands-on involvement with the natural world. It does us good, it makes us more aware of our Creator, and it also means we are less likely to misuse and exploit God's creation.

My wife has taken over an allotment in recent years, initially as a way of providing our family with cheap, local, ethically sourced food. Somewhat to her surprise, it has also deepened her relationship with God. As she waters young seedlings, she prays for friends who are new to faith. As she pulls up weeds, she examines her own conscience and prays for forgiveness and cleansing. In the planting and waiting and harvesting, she has become more aware of the way God's work and ours blend together: we may plant and enjoy the good food, but God sends the sun and rain and the miracle of transforming a tiny seed into a giant marrow.

The ways in which we relate to creation will vary according to personality and circumstance. Some people love gardening, some bird-watching, some canoeing. Some people are star-gazers, while others are keen walkers. The same person may at one stage be energized by surfing and years later enjoy the more sedate pleasures of painting landscapes. The key is that we relate to creation in some positive life-enhancing way, and through it are led to worship God.

In his foreword to the report *Sharing God's Planet*,[3] the Archbishop of Canterbury, Dr Rowan Williams, writes: 'Receive the world that

God has given. Go for a walk. Get wet. Dig the earth.' These words are simple yet very profound. God has given us his world to enjoy and to look after for him. It is deep in our nature that we should connect with creation in positive ways: experiencing the elements, enjoying God's creativity and engaging in creative work ourselves. As we do so, we deepen our relationship with the earth and with the God who longs for us to worship him.

Integration of all our relationships

One of the biggest problems we have in Western culture today is the way we box our lives into various separate compartments. Work, home, family, friends, leisure, politics, shopping and faith can all be like separate rooms in a big building, with the doors blocked between the rooms. We may struggle to relate the image of what we're like at work with how we behave at church, or how we behave at home when the masks come off. We might be competitive and aggressive in one environment, but present ourselves as gentle and loving in another. Our language in church may be rather different from our language behind the steering wheel of our cars! I'm sure you could add examples from your own experience of how your life becomes fragmented and disintegrated. It sometimes feels to me as if life is a bit like a circus performer trying to keep a large collection of plates spinning at the same time. It's only a matter of time before they all come crashing down.

This lack of integration is seen very clearly in our relationship with the natural world. We may believe that the earth is the Lord's, but it's easy for this to stay in the room labelled 'faith' and never cross over into the rooms labelled 'work', 'lifestyle', 'politics', and so on. We may worship God with our lips, but our lives (and I speak for myself here too) often give the message that we're worshipping our careers, houses, cars, bank accounts, our own bodies, and ultimately ourselves. We may proclaim that 'Jesus is Lord' but still have secret rooms in our lives which we keep firmly locked. In my experience, our whole Western culture has effectively shut Jesus out of being Lord of our relationship with material possessions and the natural world. As a result, we may be richer in our pockets, but much poorer in our relationships with God, one another and creation.

Yet God does not want it to be this way. He wants us to live lives that are connected, balanced and meaningful, lives of integrity and integration. The secret is actually very simple: put God at the centre. If Jesus is the one in whom 'all things hold together' (Colossians 1:17) – the centre of the whole universe – then all these fragmented bits of our lives will only make sense if we put him right in the middle and do not shut him out of certain areas. Letting Jesus be Lord in every part of our lives and of his creation is right at the heart of worship. Unless we do this, our lives will always be pulling us in different directions.

If Jesus is the one in whom 'all things hold together' (Colossians 1:17) – the centre of the whole universe – then all these fragmented bits of our lives will only make sense if we put him right in the middle and do not shut him out of certain areas.

In practical terms, worshipping God in an integrated way means looking at every area of our lives, asking Jesus to be Lord in each area, and taking practical steps to keep him at the centre. We cannot live our lives in isolation. Our attitudes at work will spill over into our relationships at home and at church. The secret, or hidden, rooms of our lives cannot remain that way for ever because they touch on all the other areas. Living in a way that damages the earth and the poor will damage us too, psychologically and spiritually, and eventually economically and ecologically as well.

I find it helpful to see the whole earth as a house: our home, which belongs to God, but of which we are resident caretakers. The Greek word for 'home' used in the Bible (*oikos*) is the root word for both 'ecology' (the science of relationships within our earthly home) and 'economy' (not just money, but the use of all the resources we find in our home). If we are to live lives of integrity where what we believe and say matches what we do, then Jesus – the Creator, Sustainer and Redeemer – must be Lord of our home.

Because we live in a culture that continually pulls us in competing directions, it is an ongoing process to ensure we worship Jesus in every part of our lives. In Romans 12:1, Paul addresses this task, urging us to offer ourselves 'as living sacrifices, holy and pleasing to God – this is your spiritual act of worship'. The picture of a living sacrifice is a striking one, conjuring up an image of an altar with a live sheep or goat on it. Any live animal will keep trying to jump off the altar. Similarly, we are tempted to grab back lordship over different areas of our lives and so must keep on auditing our lives and lifestyles to ensure that Jesus really is at the centre.

Prayer for God's kingdom

The Lord's Prayer is the best-known prayer ever composed, yet we easily miss its radical message. Jesus teaches us to pray:

> Your kingdom come,
> your will be done
>   on earth as it is in heaven.
> (Matthew 6:10)

We are encouraged to ask that God's kingdom rule be found not only in a far-away heaven or in the distant future, but right here and now on the earth. We have seen (in chapter five) that the kingdom of God is 'now' and also 'not yet'. It is not a physical country but rather the restored harmony of God's perfect rule breaking into our current reality and giving us a taste of heaven. This involves both practical work and fervent prayer. Sometimes Christians have seen it as one or the other: either believing we can bring in God's kingdom by political and social change, or believing that only God can change things, so all we can do is pray. The former forgets that the relationship which most needs healing is that between us and God, and without prayer we will never see real change. The latter forgets that Jesus was a man of action as well as words, spending hours in prayer, preaching the good news, and demonstrating the kingdom with powerful signs.

It has been my experience that many Christians who catch a vision for environmental work fall into the trap of action without prayer. I know of whole movements of activist Christians that have become shipwrecked on the rocks of spiritual dryness because they were not rooted in a relationship of prayer. Prayer is the lifeblood of worship, and just as a body without a living blood supply is dead, so too is worship without prayer.

Do you pray about the earth? Listening to the prayers in hundreds of churches around the UK, I notice that even when the theme of the service is creation, it is amazing how much the prayers focus only on people. Of course it is right to pray for people, but it is also right to pray for God's world and for his kingdom to be established here on earth. Just as there are prayers for the sick at every service in most churches, so perhaps there should be prayers for creation at every service.

When we see pictures of melting ice-caps, disappearing forests, bulging landfill sites, polluted seas or dying wildlife, they should move us to prayer. We should be weeping with God's Spirit at the groaning of creation, and pleading for God's mercy. I have been present in churches during times of prayer when a curtain of illusion and denial has been lifted and God has shown people just how he feels about this world and the impact our lifestyles are having on it. It has been a most powerful life-changing experience because it goes right to the centre of our being. When we begin to understand how God feels about creation, it becomes a motivating force changing hearts, minds and wills. It can lead on to a transformed world.

As we saw in chapter five, Christians have a hopeful vision of the future: a time when God's kingdom rule will once again be established throughout creation. In our prayer lives, we can hold together that vision of a transformed world with the agony of how things are now. Only then will we become transformed people who pray and work to see God's kingdom come and God's will done on earth as in heaven.

Questions

1. Using the WORSHIP outline on page 99, think through which of these areas need most attention in your own worship, and that of any church or fellowship you belong to.
2. Do you see the whole of life, including shopping and lifestyle choices, as worship? Are there some practical ways that can help you worship Jesus in the midst of everyday life?
3. Do you pray about the earth? Think of some environmental issues, local or global, that concern or upset you, and take these to God in prayer. If you (and your church) have a pattern of prayer, try to include prayers for the whole of creation on a regular basis.

8 Living it out: Lifestyle as if creation matters

Jesus, and the prophets before him, hated hypocrisy. Jesus compares religious people whose lives fail to match the standards they teach to whitewashed tombs: smart on the outside but full of rotting death. Today the media too hate hypocrisy and love nothing more than to bring down the powerful and famous when they say one thing and do another. Surveys of those who reject Christian faith often reveal that perceived hypocrisy amongst Christians is one of the things they find most off-putting. The famous Indian poet and playwright Rabindranath Tagore was once asked why, despite a deep admiration for Jesus, he never considered becoming a Christian. He replied, 'On that day when we see Jesus Christ living out his life in you, on that day we Hindus will flock to your Christ even as doves flock to their feeding ground.'[1]

For many people throughout the world today, Christianity is very closely associated with the consumer culture of the West. Although most of the world's Christians actually live in Asia, Africa and South America, the 'Christian nations' of Europe and North America are the most vocal and powerful. Although many believing Christians (myself included) would want to say that Europe and North America are in many ways no longer Christian nations, but deeply secular, one question keeps haunting me.

If somebody were to look at a street in any Western country – maybe your street – and try to spot the Christians, how easily would they find them? Apart from counting those who go to church on a Sunday, how different do Christian lifestyles look from the secular materialistic lifestyles of their neighbours? Can you spot Christians by the cars they drive (not just the bumper stickers), the contents of their shopping trolleys or the amount of waste they send to landfill? If we really believe that this earth is God's and not ours, then the lifestyles of Christians ought to be radically different from many of our neighbours.

If we are to worship God with heart, soul, mind and strength and love our neighbours as ourselves (Luke 10:27), then we need to change our lifestyles radically. At present, the average Briton uses such a large amount of the earth's resources that we would need more than three planet Earths if everybody in the world wanted to live the same way. This is both an issue of justice for the world's poor and an issue of worship, as this excessive consumerism is actually an idolatry of greed, pure spiritual cholesterol.

I am very conscious that this does not make comfortable reading. Changing our lifestyles is one of the hardest things to do, as my family and I know from experience. Most of us, tied to work and mortgages, with dependent children or parents, cannot change everything overnight. Those who are starting out and are still students or single, along with those who are retired with good health, can probably make the greatest changes quickest. Some, particularly those of the generation who grew up with post-war hardship and have always lived frugally, may have fewest changes to make. Others may take much longer and find it much harder. However, I would suggest we *all* need to make a complete, radical and honest audit of our lifestyles, their impact on the poor and on the planet, and then ask God to pinpoint where we should start making changes.

In my family, this has been a slow process that will be ongoing for the rest of our lives. I certainly don't want to pretend we'd win a 'Britain's Greenest Family' award, or that we are a model for everybody to follow. We have simply tried to put our faith into practice, and as we have become more aware of certain issues, we've sought to make changes. Because we're a family, it's been a process of joint discussion and decisions, with our children playing a full part, sometimes challenging us to go faster, and at other times reluctant to give up their favourite luxuries.

Since 2001, we have been working with the Christian environmental charity A Rocha,[2] and this has meant that we have been part of a global network of Christians who are seeking to be true to the Bible in caring for creation. We have discovered there are many others who are also on a journey of changing their lifestyles, and wanting a biblically based lifestyle commitment that would reflect A Rocha's values wherever they lived. This has led us to set up an initiative called 'Living Lightly 24:1'[3] lightening the impact our lifestyles have on the planet all day, every day. The 24:1 also refers to Psalm 24:1: 'The earth is the LORD's and everything in it.'

Living lightly

There are three core elements to 'Living Lightly 24:1': commitment, challenge and community.

The **24:1 Commitment**: The earth is the Lord's and everything in it.
Believing that this is God's world, entrusted to our responsible use and care, and that living sustainably is part of Christian worship and mission, I commit myself to:

- delighting in and worshipping God for the wonders of creation
- rethinking and, where necessary, repenting of beliefs, attitudes and lifestyle in the light of the Bible's teaching
- following the Bible's teaching even when it is counter-cultural
- living lightly in using resources as a matter of justice and worship
- supporting A Rocha as I am able

The **24:1 Challenge**: Living lightly in God's world.
Behaving differently through:

- examining and changing my values, choices and lifestyle decisions

- seeking where possible to take one new practical step towards living lightly every three months
- living with lightness, not heaviness in this wonderful, fragile world, and keeping God's 24:1 sabbath day of rest for all creation

The **24:1 Community**: Caring for God's world together.

Belonging in a renewed way to God, to my local community, to the place where God has put me, and to the family of creation:

- I recognize that, amidst a culture that proclaims freedom and individual independence, God has created me to belong in community
- I choose to belong to A Rocha's 'Living Lightly 24:1' community
- I will seek to find and join with others in modelling a sustainable way of living as followers of Jesus Christ

Believing and belonging before behaving

There are now hundreds of books, campaigns and websites that can help you with practical lifestyle tips. The difference with the 'Living Lightly 24:1' approach is that it is not only about behaviour. *Behaving* differently is sandwiched between *believing* differently and *belonging* differently. All three are essential.

Believing differently, the subject of this book, means looking at 'Why?' in a deeper way. Why should we change our lifestyles? Why should we look after the planet? Are there better reasons than self-interest? Unless lifestyle change stems from a relationship with God, there is a danger that it can simply become a new kind of legalistic religion. We should not seek to live lightly out of duty, fear or guilt but out of love: love for our neighbours, love for our fellow creatures, love for future generations, and at the deepest level of all love for God. Without this, we may reduce our carbon footprint only to find that we have become self-righteous miseries who think we're better than our neighbours!

There is a real danger of changing from one kind of hypocrite, the Christian whose lifestyle fails to reflect their belief that it's God's world, to another kind of hypocrite equally condemned by Jesus, the person who thinks they are better than those around them: 'Greener than thou' rather than 'Holier than thou'. Jesus reserved some of his harshest language for some of the most ethical, morally correct people of his day, the Pharisees, because they thought they were morally superior to anybody else. The safest way to avoid the danger of being an 'eco-Pharisee' today is to keep a simple, humble, childlike relationship of trust in, and dependence on, Jesus.

> We should not seek to live lightly out of duty, fear or guilt but out of love: love for our neighbours, love for our fellow creatures, love for future generations, and at the deepest level of all love for God.

Belonging differently is also extremely important. It is nearly impossible to change your lifestyle radically in isolation. You need a group, even a very small group, of others who are on the same journey. It's partly that we all need people to encourage us and keep us accountable. It's partly because living lightly is about sharing, realizing we don't all need hedge-trimmers or waffle-makers, but can share and swap. It's partly because the trend towards one-person households is environmentally (never mind socially) very damaging. The simplest way of reducing your carbon footprint is to share a home, thereby sharing heating, cooking, travelling, lighting and consumer goods with several other people.

Belonging differently matters at a deeper level too. In a fragmented, rootless, disintegrating world, we need people who will commit themselves to one another and to local places and bring about transformation that begins locally but produces an attractively infectious vision of living differently.

Can you envisage a time when local churches will become pinpoints of light in a time of growing environmental darkness? This is my dream: lifestyles that model both environmental sustainability and living with lightness and joy in God's world. Harmonious relationships with one another, God and creation could be the key to transforming our whole culture. People are increasingly disillusioned with the empty rewards of consumerism and escapism. Millions are seeking a more authentic way of life that connects with nature and makes sense of the environmental crisis, a way that uses the benefits of technology carefully and wisely, relates to the search for spiritual reality, values people in their brokenness, and offers the chance of healing and transformed relationships.

Within A Rocha we have seen small hints of this, with people on a spiritual journey often turning up at the centres which form the heart of A Rocha's practical environmental work, and finding, to their surprise, that Christianity does indeed have something relevant to offer. As Christians today rediscover the power of community, lived relationally and sustainably, we may find that we have also stumbled across the key to reconnecting with a culture that thinks Jesus has nothing to offer, yet has never been hungrier for what he really offers.

Behaving differently

In terms of practicalities, Living Lightly 24:1 is organized around fifteen different dimensions of our lifestyles:

Action:	campaigning on behalf of the poor and voiceless, both human and non-human
Church:	transforming our places of worship and the people who meet there
Food:	eating gratefully, ethically, locally, simply and joyfully
Friends:	sharing our values through our relationships
Garden:	earthing our faith by working with creation in a hands-on way
House:	homes that are sustainable, energy-efficient and community-building
Leisure:	resting and just 'being', respecting the givenness of God's world
Money:	investing in God's kingdom through our spending, saving and attitudes
Nature:	rediscovering our place through enjoying, studying and caring for creation

Quirky:	remembering that today's crazy idea is often tomorrow's prophetic action
Seasonal:	relearning God's creation rhythms, through festivals, food and fun
Shopping:	buying into values that last, rather than stuff that doesn't
Travel:	questioning the need for speed and being responsible for our impact
Waste:	the '3 Rs' – reduce, re-use, recycle, plus the fourth . . . refuse it in the first place!
Work:	sharing our sustainable values in the workplace

There isn't space here to go into all of these dimensions in detail, and it might well depress you if I did! Rather, I want to share something of what we have found helpful, the practical experiences that lie behind 'Living Lightly 24:1' and have formed part of our journey.

One step at a time

One danger with looking at your lifestyle is that you quickly realize just how much you need to change. It's all too tempting to give up before you start. When I began to think things through seriously, it seemed as if there was an impossible mountain ahead of me, that I needed to go back to the very foundations my life was based on and start again from scratch. Yet I still had to go to work, feed and clothe my family, travel from A to B, and live within a culture from which I felt more and more alienated.

Waste: the '3 Rs' – reduce, re-use, recycle, plus the fourth . . . refuse it in the first place!

It was a real release for me to understand that my responsibility was not to do everything at once, but to turn to God and ask, 'Where do you want me to start?' Together with my wife, I asked God what *one* thing he wanted us to start with. The answer was a bit surprising: 'Nappies'! We were expecting our first child and read horrifying statistics about the percentage of landfill sites that were filled with non-biodegradable nappies.

My wife did some web-based research and we discovered that we could get shaped, washable, cloth nappies with waterproof covers. They cost a couple of hundred pounds to buy, but over their lifetime they have saved us several thousand. Not only were they passed on to each of our children in turn, but, they were borrowed by others in between! More importantly, we were not throwing out sack-loads of nappies each week. We used biodegradable disposable liners (so they never got too stained), soaked them with bicarbonate of soda, machine-washed them at a low temperature, and whenever possible dried them outdoors. Even nappies that looked indelibly stained came up gleaming white after an hour or two of being bleached in the sun (yes, even the grey cold sun of an English winter!).

Some people questioned whether it was really more 'eco-friendly' to wash nappies, as the washing and drying might contribute extra CO_2 to the atmosphere. However, the latest research is very clear that washable nappies laundered at the right temperature, dried wherever possible without a tumble-dryer, and used on more than one child, are far more eco-friendly than any type of disposable.[4] Moreover, CO_2 is not the only issue: the mountains of non-recycled waste we all create are in themselves a major environmental problem, with 3 billion disposable nappies used in the UK each year and 90% going into landfill sites.

Of course, anticipating the obvious question, it was more work than using disposables. But, think of the advantages: we saved money, created far less waste, and were in a position to share our reasons for making this choice wherever we went. And actually, with modern washing machines, it wasn't massively more time-consuming. For us it was, perhaps most importantly, a decision that changed our way of thinking and started us on the journey that we're still on.

For you it may well not be nappies, but why not ask God to put his finger on one area of your life where the change needs to start? You will be amazed where this journey will lead.

You are what you eat

'Food, glorious food' – I love it! The problem is that people in the West love it too much. In the UK we live in a society where about 30% of food is thrown away. We have a growing childhood obesity problem, yet we live in a world where millions are starving. Like the whole issue of living simply, 'food' is a vast and bewildering subject. Issues arise around where it comes from (food miles, fair-trade), how it's produced (organically or not, additives and processing, use of pesticides and fertilisers), the conditions in which animals are kept, how it gets to us (packaging, air and road freight), and the balance of what we're actually eating (healthy and seasonal).

As a family, we began in very small ways. We grew some herbs in the garden (neither of us ever having gardened before) and began to revel in the spontaneous pleasure of adding some chives or mint to a salad. My wife then decided to grow a little more, enlisting the children's help in planting some tomatoes and carrots. Not only did we feel good about the vegetables (they had no food-miles, and were organic and seasonal), but they were also the most delicious ones we had ever tasted. My wife made the huge leap of taking on an allotment and, with lots of advice from friends and some web-based research, we found ourselves eating lots of home-grown squashes, sprouts, beans, salad-leaves and even raspberries and strawberries. It was a big investment of time, and hard, physical work, but also incredibly rewarding, not just in terms of good food, but in terms of relationships within and beyond our family.

At the same time, we began to look at the balance of what we ate, and realized we were eating a lot of meat without thinking very hard about where it came from. I did some Bible study on animal welfare and was genuinely shocked at what I found. Not only are animals blessed as 'very good' in creation and included in God's saving covenant at the time of Noah, they are also repeatedly and specifically protected against abuse in the Old Testament.[5] The book of Proverbs clearly states, 'A righteous man cares for the needs of his animal, but the kindest acts of the wicked are cruel' (Proverbs 12:10). If God cares so much about animal welfare and if mistreating animals reflects on my relationship with God, then my eating habits had to change.

I had heard facts and figures about animal welfare and the meat industry, but a little research revealed a number of shocking issues. Many chickens and pigs in particular are kept in very inhumane conditions, crammed into small spaces, and deprived of everything except excessive amounts of food. Cattle-ranching to feed the appetites of Europe and North America is a major cause of the destruction of the Amazonian rainforest, where it is estimated that a pound of beef 'costs' 200 square feet of rainforest.[6] Moreover, much beef and lamb travels from the far side of the globe, adding to carbon dioxide emissions.

What should we do? One of my daughters decided to go vegetarian completely. The rest of the family decided we would cut down on the percentage of meat in our diet, eating it only two or three days a week, and only meat that was as local as possible, and in the case of poultry and pork, free-range or organic. The result? Our 'veggie' daughter has now decided that she's content with our 'happy meat' policy and so we can all eat together. We are also now eating more healthily and enjoying it far more.

Of course, eating ethically is more expensive, but despite being a growing family and having an income substantially below the national average, we haven't noticed a major effect on our finances, because we are eating much less meat, have begun to grow some of our own produce, and are also constantly on the look-out for special offers on ethical food.

As I've said, our lives are not meant to be a prescription for everybody else's. We know there's more we could do, and not everybody will agree with every step we've taken. Some may decide to go completely vegetarian. We didn't because Jesus wasn't a vegetarian, and we felt it important to support those farmers who are trying to produce meat in an ethical way. Some people may be able to raise their own chickens, sheep or pigs, and I have growing sympathy for those who say you should never eat something that you are not prepared to see being killed.

Before leaving the subject of food (and there's so much left unsaid), I want to mention the importance of enjoying what we eat. Mealtimes should be celebrations of God's creation, not just refuelling stops in a twenty-four-hour race. It is no coincidence that festivals in every culture worldwide include food. As a family, we have reinstated 'proper' meal-times with no television, phones, MP3 players or other distractions, where we actually sit around a table, give thanks to God before we eat, and talk to one another.

Occasionally they will be really special meals: birthday breakfasts with croissants and balloons, or spontaneous dinners on a national theme (Mexican, Indian or Lebanese) with menu cards, candles and even fancy dress. We may discuss where our food has come from, imagining the field it grew in and the people who grew it, while thanking God for providing it and sustaining us. At least twice a week, we deliberately eat a very simple meal of rice and dhal (spiced lentil stew) to identify with the many people in our world for whom this is a staple diet, all day, every day.

Waste not, want not!

Closely linked to food is the subject of waste, as a huge proportion of waste is from food and its packaging. One of the first things we did was to stop buying individual yogurt pots with 150 ml of yogurt surrounded with plastic. Instead we bought larger 500 ml or one-litre containers and shared the contents. After a while we went a step further and acquired a yogurt-maker. All it took was a litre of long-life milk and a spoonful of live yogurt, and we had delicious home-made yogurt to which we could add our own fruit or flavourings. Again, one small step at a time, and we're not so strict that we won't buy yogurts if we're over-busy or have forgotten to make some. It's important to keep a sense of perspective and a sense of humour.

Perhaps the single most effective step I have taken in terms of avoiding waste is to try to pray every time I put something in the bin. As I do so, I thank God for the natural resources that have created the item, and reflect on whether I have been a good steward. Often my prayers end up as guilty confessions, as I have to admit my careless wastefulness again. Did I need to buy this? Did I need the item with all this packaging? Sometimes my anguish is not personal so much as frustration with a culture that dictates I can't buy things that aren't covered in polystyrene and plastic, and I simply say sorry to God for what we as the human race have done in taking his world for granted.

Seeing waste as a spiritual issue has caused me to go back to the start of the process: buying things in the first place. Now, before I buy something, I always try to ask whether I really need it. Shopping today has become a major leisure pursuit rather than a way of getting what we need. People shop in order to feel good rather than to feed or clothe themselves. Yet all our 'stuff' can simply become a way of shutting out the real world or blocking out God's uncomfortable voice. With the 'Living Lightly 24:1' commitment, we send everybody who signs up a credit-card-sized reminder to keep in their wallets so that every time they reach for their cards to buy something, they're reminded that the earth is the Lord's, and not just there for our consumption.

Refusing new things, reducing our shopping, and re-using rather than replacing are all good steps, but that still leaves recycling. Some years ago we did not recycle anything, and our weekly waste went into three or four large black bin-liners. Now, even with a larger family, we rarely put out a full bag each week. Once again, this change hasn't happened all at once. We began with recycling bottles and cans in the days when there was no doorstep collection and we had to transport them all to the supermarket car park. Now, as local authorities add to their services under the threat of fines from the European Union, it's become much easier to recycle in most parts of the UK. In our home, we simply have a series of bins in our kitchen, and just outside the back door two others for compost and other food waste. Even the kids have separate bins in their rooms for paper and other rubbish, so they learn young. It's not difficult!

Travelling ethically

Travel brings freedom, but it can also enslave us. We see the world, but may lose sight of who we are. As our motorways become traffic jams, and we spend longer in departure lounges than in the air, perhaps we should ask if we are running away from something?

For many, if not most of us, breaking the habit of always using the car even for short journeys is a good place to start. Our family lives almost a mile from the primary school our daughters attend and my wife and I nearly always walk them to and from school. On the way, we pass people getting into their cars to drive to the same school. Some of them are on one-way sidestreets, so need to drive over a mile in order to get to a school that is only 300 metres from their house! When they arrive at the school, they often find that there's nowhere to park, so they may have to drive back to the street they started on to find a space. What on earth is going on here? People say they feel safer driving to school in a car, but watching the cars double and triple parked outside the gates, with children dodging in and out between them, breathing in the fumes as their parents sit with engines running, I wonder what planet they are on. It's certainly not the planet God made us to live on.

Of course, it's not just the 'school run'. Two years ago I took up cycling again after a long lapse, and I have enjoyed its multiple benefits: I am physically fitter, see more of my local community, and in London regularly get to places more quickly than I used to by car. It may be that your health or location makes cycling impractical, in which case it is worth asking what the public transport options are. In a large city, these are likely to be quite frequent and at least as time-efficient as our over-clogged roads. When travelling longer distances, train travel is generally much better than the media often portrays it to be. If you book tickets at least a week in advance, they can be incredibly cheap. I decided to take my oldest daughter for a couple of days of winter walking in Scotland in February, and we managed to get berths on the overnight sleeper from London to Inverness for £19 each way.

Air travel is widely known to be the biggest single problem in terms of travel emissions. Not only do planes produce a huge amount of carbon dioxide, especially when taking off and landing, but they also release other climate-changing gases. It is estimated that a single return flight from London to New York is the equivalent of a whole year's average mileage in an average car.[7] Short flights are proportionately the worst environmentally, so a growing number of organizations, A Rocha included, have made it policy not to allow internal flights within the UK. As an international movement that believes relationships are vital, we do still use flying where it is unavoidable and where the benefits outweigh the costs, but we agonize over this and also offset all our travel.[8]

As a family, we've decided to holiday mostly in the UK rather than take advantage of artificially cheap European flights, subsidised by a lack of tax on aviation fuel. When I was asked to speak at a conference in Austria, and invited to bring the family along, we agonized over the best method of travel. The plane was relatively cheap but environmentally terrible, while the train (just after Christmas) was prohibitively expensive and, somewhat to our surprise, only slightly better than driving the whole way in our trusted family diesel. So we drove to Austria, leaving on Christmas Day and taking two days each way for the journey. The children rated being cooped up in a car for 48 hours as part of one of their best holidays ever! They got to see parts of France, Belgium, Luxembourg, Germany and Austria, and we did our best to make it all interesting with snippets of language, food, music and history from the places we went through. If we had flown, they would have had no concept of Europe, the distances and the cultures they had crossed, but would simply have stepped straight from London into Austria with no connection.

Perhaps the biggest problem with easy travel is the loss of a sense of place and distance. Living lightly in God's world means looking at how we travel, as well us bringing us to the point of travelling less. It means putting down roots and getting to know the places in which God has planted us.

Coming home to roost

The houses we live in and the energy they consume are our biggest single contribution to global warming, averaging around 40% of our total ecological footprint. There are now dozens of websites, including an official government one,[9] where you can look at your home, work out your carbon footprint, and get advice on how to do better. Such sites are very practical and helped me realize, for example, that the single, biggest step we could take as a family was to improve insulation throughout our home. This is far cheaper and simpler than putting in solar panels, new boilers or wind generators. Although it rarely receives as much publicity, it can reduce your home's carbon footprint significantly.

However, most websites won't help you with a much deeper issue: why we feel as we do about our houses. Our homes express our values and personalities: which rooms we think are the most important, whether we are obsessed with tidiness or not, how many gadgets we have. If you want to be really challenged, why not try this? Walk around your home, inside and out, structure and furnishings, and prayerfully ask what Jesus, accompanied by an African Christian, would make of each area. It is almost a shocking suggestion, because our homes are often our last defence against reality. They can become an insulated bubble where we create our own version of how things ought to be. They can be the last area where we allow Jesus to be Lord of the whole of our lives. Yet it is incredibly releasing to allow God to challenge us gently about our homes and their contents. This suggestion is not intended to burden you with extra guilt, but rather to help lift the burden of worry and care that materialism often imposes. We don't have to keep up with our neighbours. We don't have to buy all the latest gadgets we see advertised. We don't have to be slaves to fashion. We are free to live joyfully and simply in God's world.

Celebrating simplicity

Simplicity is an inner value, not just an outer one. It is about getting rid of our mental, not just our physical, clutter. I cannot pretend I find this easy. Life today is like a leaky boat; it fills up as soon as you empty it. Yet as a family we've found that a few fairly simple steps have begun to have far-reaching implications.

During Holy Week this year, we decided that we would try and live without electricity or gas from Good Friday to Easter Sunday. It didn't seem like a world-changing decision: about sixty hours without all the usual comforts, yet it has affected us quite deeply. We turned off all our lights and electrical appliances (except for the freezer), and our central heating and hot water. We decided to cook outdoors on an open fire, using scrap wood and (sustainably sourced) charcoal. No doubt this was less energy efficient than using gas or a microwave, but this wasn't only about saving carbon, it was about changing the way we think about our lifestyles. The younger children went to bed as soon as it got dark and we stayed up by candlelight reading and chatting with the older ones: no TV, no music, no emails or computer games.

In the morning, no alarm clock! My wife loves an early morning cup of tea, and we had not thought of saving hot water in a flask. So we had to get dressed, go outside, build a fire and wait for the water to boil. As the weekend went on, our meals became community events, with friends and passers-by popping in and joining us for a simple vegetable stew or toasted marshmallows. There were difficulties: everything took so long, we had to put on warmer clothes, and shaving and washing were hard work. Yet it made us all think, identifying with the darkness and silence Jesus went through between Good Friday and Easter Sunday in a new way, and entering into the joy and light of Easter. It also forced us to consider what life is like for all those who regularly live without power, gathering firewood to cook or boil water. We didn't try showering or bathing in our two and a half days, and were a bit horrified when our enthusiastic five year old announced to our church that she wanted us to do the same thing for the whole of Lent the following year!

Of course, our experience was both temporary and only a tiny taste of what most people in the world have to put up with – we still had running water and indoor flushing toilets. We had ready-to-eat food in the cupboards, fresh bread from the shop, and the knowledge that this was only a very brief interlude in our normal lives. But it actually made us both more thankful for modern technology and all its benefits, and more careful in how we use it.

The real surprise to me was that our 'Easter electric fast' was also highly enjoyable. It became a celebration of simplicity, a talking-point with others, a treasured memory for our children, and a time of working together instead of all doing our own thing. It reminded me of another occasion when we told all our family and friends that we did not want them to buy us Christmas presents, but that we hoped to make some simple gifts ourselves. We had become sick of the commercialised glitz and shallow materialism of Christmas, and wanted something simpler and more authentic. Presents, especially for children, can become no more than an attempt to buy affection or a pandering to our advertising industry that wants everything to be bigger and better than last year. Many presents, composed of oil-based plastics, are made by underpaid workers in Far-Eastern factories, padded with vast amounts of unnecessary and unrecyclable packaging, and in due course broken or discarded within days of being received. Why? As a family, we created hand-made cards, using our own under-age workforce (!), and spent evenings making home-made organic chocolate fudge. The one commodity that was costly was time, yet it only entailed switching off the TV and doing something together as a family for a couple of evenings.

Simplicity is first and foremost a state of mind which then expresses itself in our lifestyles. It is not about turning the clock back to the myth of a pre-industrial ideal. Rather, it is about being focused on what really matters, seeing the wood for the trees. In biblical terms it is about seeking first God's kingdom and his righteousness and letting 'all these things' – possessions, food, money, clothes – take care of themselves (Matthew 6:33). It means realizing that relationships are always, *always*, more important than tasks, deadlines and material things. In practical terms, it can begin with small but radical steps, such as walking instead of driving for a week, turning off the power, or giving up television or email for a short period.

What I keep discovering, as with my family I try to cultivate a simpler lifestyle, is the sheer joy of letting go of stuff! Living a lifestyle that cares for creation and the poor has ceased to be about guilt and duty and has become a discipline of delight. It is a discipline because our culture keeps pulling us back into resource-intensive, consumer-driven, greedy lifestyles, but it is also about becoming more carefree: free of the care and stress of a competitive culture, and increasingly conscious of our dependence on others, on creation, and on God.

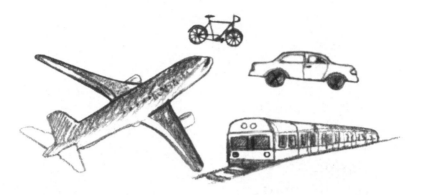

Questions

1. 'One step at a time.' What is your next practical step in your lifestyle change?

2. The chapter talks about 'behaving, believing and belonging'. Behaviour and belief are addressed in this book and many other places, but what about 'belonging'? Are there others you know who could join you in exploring living lightly?

3. What would you want your lifestyle to look like in five years time? Is this a sustainable vision that honours God, includes justice for the poor and cares for creation?

9 Living it out: Mission as if creation matters

In chapter one I raised a question which I have often been asked: 'Shouldn't we just be focusing on evangelism, rather than worrying about the planet?' It is a good question and one which I take very seriously. I am passionate about the *evangel*, literally the 'good news' of Jesus Christ, and about communicating Jesus to everybody I meet in every appropriate way. I believe that it is only through Jesus that we can have hope, both for ourselves and for this world.

The great commission

It would be a huge distortion of the Bible to say that 'evangelism' (in the sense of bringing individuals the saving news of Jesus) is the only thing that God cares about or that Jesus told us to do. When the risen Jesus sent his disciples out into the world, he issued what is often called 'the great commission'. It appears in slightly different versions in each of the four Gospels and is most often quoted from Matthew 28:19–20:

> 'Therefore go and make disciples of all nations, baptizing
> them in the name of the Father and of the Son and of the

Holy Spirit, and teaching them to obey everything I have
commanded you.'

There are a couple of important things to note about these famous
words. First, they are not a command to go and make 'converts', but
to make 'disciples'. Evangelism on its own is about leading people
to Jesus. Discipleship is about leading people on a lifelong journey
with Jesus. It's about recognizing that conversion is only the
beginning of a lifetime of learning 'everything I have commanded
you'. Discipleship, as well as evangelism, is the mission Jesus entrusts
to his followers and the mission still we have today. 'Everything I
have commanded' includes issues of justice and the environment.

Secondly, we must balance Matthew 28 with Mark's version of the
great commission, which gives us a wider vision. Here Jesus says: 'Go
into all the world and preach the good news to all creation' (Mark
16:15). Scholars have often ignored this verse, reminding us that it is
not in all the oldest manuscripts. However, it is included in our Bibles
as God's inspired word, and it is possible that it is ignored because its
implication is so radical. Our mission, our God-given great commission, includes the whole of God's creation, not just people. We could
joke about how to preach to a goat or a garden, but that would miss
the point. This is not only about preaching with words, but about
how we communicate Jesus' message in our relationships, attitudes
and lifestyles. What is the good news for a rainforest that is being
chopped down to feed our meat-hungry lifestyles? What is the good
news for those who face the spread of deserts and failing crops as a
result? What is the good news for creatures that God lovingly made
but are now driven towards extinction? What is the good news for
the world's climate systems as they become thrown off course by our
polluting lifestyles?

I believe there is good news for all these situations, and it is found
in Jesus. It is the good news of the cross and resurrection. It is the
good news that God is committed to sustaining and renewing
the creation he made in love. It is the good news that sinful, greedy,
polluting human beings can be transformed inwardly in their
relationship with God, and outwardly in their relationship with
others and the planet. This is the good news that our world so
desperately needs, our mission at a time of ecological crisis.

Thirdly, the great commission given us by Jesus in the New Testament must be held alongside the very first great commission God gave us at the start of the Bible. In Genesis 1, God's very first words to human beings are about ruling over and caring for creation: the fish, the birds and all the other living creatures, for God's sake. This is, if you like, a universal job description of what it means to be human. To the question 'Why are we here?', the ultimate answer has to be: 'To worship and serve God.' The first element of that worship and service that the Bible talks about is creation care.

I don't in any way want to underplay Jesus' words in Matthew 28 or the call to engage in evangelism, but Christians have often emphasized this to the point of ignoring our mission to the wider creation. Mission has been well described as 'using the whole church to bring the whole gospel to the whole world'.[1] The whole church includes every Christian without exception. The whole gospel means the good news of Jesus as it applies to every dimension: spiritual, physical, social and environmental. The whole world includes not just people of every nation, but the whole creation, made by and for Jesus, finding its purpose in Jesus, and able to enjoy its freedom because of what Jesus has done.

The scope of mission

In the past, many Christians have seen our God-given mission as, in essence, rescuing dying people from a doomed planet as this world runs out of control like a runaway train. Jesus has provided an escape plan, the church, a bit like a railway carriage that is going to be uncoupled and taken to safety when the rest of the train heads over the cliff to destruction. Mission therefore is a race against time to get people into the right carriage before it is too late.

The Bible's radical message is that God's mission is far bigger and far more exciting than this. Although the train, the world, and everybody on it, are indeed running out of control towards destruction, there is still hope. God's rescue mission is not for a few passengers only, but for the train itself. To return to biblical language, it is not only people whom God made, loves, cares for and brought back into relationship with himself through Jesus, but the whole of creation.

Our task as human beings is to bring the good news of Jesus to all of creation in both word and deed. This is not about *our* mission to save people like us. It is *God's* mission of good news for the whole created order. The biblical evidence is that mission is wider than we've often allowed. Humans may be the key agents of mission in the new creation that Jesus has begun, but we are not the only objects of mission. Mission is ultimately not only about people but about the renewal of all things in Christ. The transformation of individuals by the good news of the gospel still is, and always will be, central, but the biblical gospel of Jesus Christ is also good news for every creature.

One helpful model to understand this wider definition of mission is the 'five marks of mission':[2]

- to proclaim the good news of the kingdom
- to teach, baptize and nurture new believers
- to respond to human need by loving service
- to seek to transform unjust structures of society
- to strive to safeguard the integrity of creation and sustain and renew the earth

The first two marks cover the traditional understanding of Jesus' words about making disciples from all nations in Matthew 28:19–20. They are essentially about evangelism and discipleship. The third and fourth marks encompass the Christian response to the suffering in the world, our mission towards the poor and marginalized, and our commitment to challenging and changing injustice. The fifth mark is creation care, our mission towards the non-human creation.

If we emphasize only one or two of these, we will always present a less than complete gospel. For instance, there are many places where evangelism and discipleship alone have been understood as the whole of mission. Here, churches may become filled with people who have come to know the forgiveness of sins that Jesus brings and learn to study the Bible, and pray and worship together. However, the society and environment around them may not be touched by the gospel. There may still be ethnic or social divisions. Racism, slavery and caste may all be tolerated. The rich may still exploit the poor, and people may exploit the environment in a selfish but unthinking way. The key reason why Christianity has been ineffect-ive in bringing total transformation is that Christians have been given an incomplete model of mission and the gospel. Sometimes this has led to terrible tragedies, where times of great spiritual revival have failed to prevent ethnic hatred and even wars, all because the gospel preached was only about 'spiritual' things and failed to touch people's relationships with one another.

Across the world, we can see the same environmentally. Chris-tians have often been criticized for allowing and even encouraging the destruction of the planet. When we have preached only a gospel about spiritual change or an 'otherworldly heaven', this is fair criticism. The whole gospel is about Jesus transforming a person's relationship with God, and their relationships with other people and the world around them.

Recently I saw a signboard in the middle of a major capital city in Africa. The sign read, 'Jesus Christ is the Lord.' In that particular country 70–80% of people attend church regularly and there are high levels of personal commitment to Christ. Yet, as I looked at the view around that sign I saw something very troubling. Behind it rose gleaming glass and steel skyscrapers, pointing to material success and a thriving economy. In front of it lay a shanty town where thousands of people live in filth and poverty. A river nearby was choked with rubbish, rotting animals and open sewers, and I could see many charcoal sellers, evidence of an illegal trade that is destroying native forests. I knew from talking to local Christian leaders that huge problems of corruption and Aids exist, within as well as beyond the churches. The country faces an environmental catastrophe, with encroaching deserts, disappearing forests and a population explosion. The question hit me like a steam train: what does it mean to say 'Jesus Christ is the Lord' in this context? Is the good news only about escaping and dreaming of heaven? No! If we really believe what the Bible says about Jesus, then he is good news for the thriving businesses and the struggling poor, the polluted river and the threatened wildlife, and those with Aids or tempted by corruption.

Thankfully, in this part of Africa an increasing number of Christian leaders are catching this wonderfully transforming biblical vision. Christians are getting involved in the slums, in challenging corrupt politicians, in practising sustainable farming, and in questioning the destruction of the environment. Yet in doing all this, they are not losing sight of the importance of the 'spiritual' gospel, as they continue to preach forgiveness of sins, worship God with enthusiasm, and underpin their work with passionate prayer. All the marks of mission are seen together, bringing about genuine transformation.

Special agents

Today, right around the world, Christians are at last rediscovering mission as if creation matters. It is happening as people look at the Bible again and discover God's passion for this planet. It is happening as people see the environmental crisis around them and ask how God feels about what we're doing to his world. It is happening as Christians seek to answer their friends' questions about what hope Christianity can offer. It is happening as increasing numbers feel frustrated with the inequalities in the world and the empty pleasures of material prosperity. It is also happening as – often quite spontaneously – people sense God's Spirit calling them into caring for creation in Jesus' name.

I have been hugely blessed to be involved in a small way in this movement that God is beginning. It is now over fifteen years since I sensed God challenging me about his world and how he felt about it. I haven't space to tell that story here. However, it led me to getting involved with A Rocha, an international Christian conservation agency committed to caring for creation in practical ways. When I first came across A Rocha, it was one small project in Portugal looking after a threatened estuary. It seemed very small and a bit quirky: how could this be mission? Yet I discovered that what was taking place through that project and the people who were working there was so special that it made me go back to the Bible and re-think my faith completely. I was enormously challenged to realize that God's very first call on us as we read the Bible is to care for our fellow-creatures for his sake. I came to realize that for Christians to spend weeks or years studying plants, molluscs or birds was not a waste of time or a distraction from 'real' mission, but part of our mission in God's world, fulfilling the commands to study, understand and care for creation. I began to see the links between caring for creation and caring for people, and how the two are inseparable.

Others have seen A Rocha's work and caught the same vision. From one project in Portugal, it has expanded over the last fifteen years to projects in nearly twenty countries.[3] This has not happened through a clever business plan or a rich backer: often the opposite as people have had to give up jobs, explain themselves to churches that didn't understand them, and raise their own financial support to work with the organization. It has been as God's Spirit has challenged people, often in quite independent and apparently random ways. What, for instance, do the following people have in common: a young couple involved in the media in Singapore, a Lutheran pastor in the Czech Republic, and a biology lecturer in Peru? Answer: they all thought they were the only Christians in the world God had called into a mission of caring for creation in Jesus' name until they stumbled across A Rocha!

We are not using environmental work as an excuse to smuggle in the 'real' spiritual gospel. Instead, we are living out and sharing all that God has called us to: our part in God's mission in his world.

Through the ministry of A Rocha in very different places around the world, we are learning all the time about the practice of mission as if creation really matters. Our projects aim to demonstrate God's love for people and the rest of creation together. They involve prayer and Bible study and talking about our faith, alongside scientific research, environmental education and sharing our lives as we explore sustainable community. If you spend time helping in an A Rocha project, you may find yourself picking up litter, counting butterflies, filing papers, washing up, teaching children or clearing nettles. All of these are mission if done as part of worshipping and serving Christ.

In responding to the biblical call to care for creation, we find that often people without a Christian faith are attracted by what they see. The evangelist Dr Rob Frost has said, 'When Christians take the earth seriously, people take the gospel seriously.'[4] This has been our experience in A Rocha. We are not using environmental work as an excuse to smuggle in the 'real' spiritual gospel. Instead, we are living out and sharing all that God has called us to: our part in God's mission in his world. This is what seems to attract those who are searching for a spiritual reality which makes sense in a world of damaged people and degraded ecosystems.

As A Rocha continues to grow, I have realized something further. This call to mission as if creation matters is not only for biologists, botanists or butterfly experts. It is for all of us. Romans 8:19 states, 'The whole creation waits in eager expectation for the sons of God to be revealed.' Who are the sons of God? They are every Christian, male and female. So, for what is creation waiting so eagerly? I believe it is waiting for us to rediscover our God-given role as its stewards and caretakers. The role that God gave Adam at the beginning of serving and preserving his creation (Genesis 2:15), in which we have failed so terribly, has been made possible again by the 'second Adam', Jesus Christ. We have been renewed and restored in Christ's image, and we can now work with God in his world for the renewal and restoration of creation. We are God's special agents in this broken and needy world; people on a mission to bring the good news to the whole creation.

This is incredibly exciting. God has made you and me his representatives on planet Earth. He has entrusted us – yes, us! – with being his special agents in repairing, renewing and caring for his creation. What an amazing privilege! What a huge responsibility! What a calling to live up to!

James Jones, the Bishop of Liverpool, puts it this way in speaking of Jesus' mission and ours: 'And what is that mission? To do God's will on earth as it is done in Heaven. His [Jesus'] prayer and ours. His mission and ours. The earthing of Heaven. The *Missio Dei*. The Mission of God.'[5]

The final challenge of this book is this: will you join in? In the Old Testament, when human sin had led to the threat of a disaster that would engulf people and the rest of creation, one man responded to God's call. His name was Noah, and he discovered that God's mission wasn't only about people but included rather a lot of other creatures too! Today we need a whole generation of Noahs, people who seek to follow their faith in God by demonstrating Jesus' saving lordship over all creation. Today we need Christians who are prepared to think differently, worship differently and live differently from those around them.

We can draw our inspiration from the biblical vision that one day God's plan is 'to bring all things in heaven and on earth together under one head, even Christ' (Ephesians 1:10). In a world where each week brings worrying news about the earth and its future, and where there is increasing despair, we are called to live out God's hope for the planet. Although this world will only be put right completely when Jesus returns and gets rid of all the sin, suffering and evil that have infected it, we are to be his advance party, welcoming signs of his rule here and now. Our task is to enable signs of God's kingdom on earth as in heaven, to be agents of change in welcoming and establishing the reality of Jesus' lordship in the whole of creation.

In the classic Bollywood film *Salaam-e-Ishq* there is a scene where an unlikely hero, a lovelorn middle-aged Delhi taxi driver, is berated by a friend for dreaming of his perfect 'Miss Right'. His reply is this: 'It's not a dream ... It's a fact of future truth.' Christian hope is no dream. Founded on the reality of a God who creates, sustains and saves, it is based on solid foundations. However bad the situation around us, hope for the planet is a fact of future truth.

As we pray and work for Jesus' world, as we seek to repair the damage we have done and as we preach the gospel to all creation, we may catch glimpses of how things will be one day. What we are part of now can only ever be the palest reflection of that day, but just imagine all the best things of this world, all the most beautiful, most inspiring, truest and loveliest things in all creation, combined with the glorious perfection of God's presence in heaven. Imagine God once more making his home amongst human beings (Revelation 21:3). Imagine creation set free from its bondage to decay, all things released from death and suffering, and earth and heaven reconciled to God. As followers of Jesus, let this be our vision, and let this be our mission.

> Joy to the world! the Lord is come!
>   Let earth receive her King;
> Let every heart prepare him room,
>   And heaven and nature sing.
> (Isaac Watts, 1719)

Questions

1. If the command to reflect God in caring for creation is 'the first great commission', how is it reflected in your priorities?
2. Do you, and does your fellowship or church, support each of the five marks of mission (page 137) through praying, giving and practical action?
3. How can we keep a full biblical understanding of mission, including creation care, without neglecting the importance of sharing the good news of Jesus with individual people? Does the work of A Rocha (see <www.arocha.org>) offer some helpful ideas?

Appendix: Planet whys

Frequently asked questions (FAQs) on Christianity and the environment

1. 'Shouldn't we be focusing on evangelism, not environment – saving souls rather than saving the earth?'
Firstly, Jesus didn't only focus on 'saving souls'! He cared for whole people in their physical and social as well as spiritual context. To Jesus, people's relationship with God could not be separated from their relationships with one another and the world around them. He taught that loving God and loving your neighbour are linked. So, healing the sick, releasing the prisoner and stilling creation's storms are all part of the 'good news' (the gospel of the kingdom of God) that Jesus taught and modelled. Evangelism ('saving souls') is a core Christian calling and people can only come into a living relationship with God when their sins are forgiven through Christ, but evangelism should not be separated from living out the whole gospel. Take the example of Noah's ark – it's about God's purposes in rescuing us from the effects of sin . . . yet it was not only 'souls' but whole people who were saved. In fact, it wasn't only people but 'every living creature upon the earth' (see Genesis 6 – 7), so perhaps God's view of what needs saving is somewhat bigger than ours has sometimes been!

Secondly, when evangelism is not only words but is accompanied by Christians showing God's care for the whole world in practical ways, it is much more powerful. When Christians have nothing to say about today's 'big issues', including the environment, it puts many people off Christianity. In contrast (to quote the late Rob Frost), 'when Christians take the earth seriously, people take the gospel seriously'.[1] This is the experience of A Rocha's projects, where for many people the Christian faith suddenly seems to make sense when they see it lived out in relationship to other people and the whole planet. So, in conclusion, it's not a case of either evangelism or saving the earth, but both the good news of salvation *and* good news for creation.

2. 'Isn't the gospel about spiritual matters, not material ones?' 'Doesn't God care about our souls, not our bodies?'

Underneath this question (which comes in many guises) is the deeply flawed idea that we can separate the 'spiritual' from the 'material' or physical. In reality, the Bible always sees human beings as a mind-body-soul unity that cannot be separated. We are whole people composed of physical, mental, emotional and spiritual capacities, not simply immortal souls encased in physical bodies. This latter idea stems not from the Bible but from Greek pagan philosophy.

The very fact of creation, that God has made a material universe and declared it 'very good' (Genesis 1:31), and that God continues to uphold, sustain and renew the creation, shows that material things do matter to God. Even more so, the coming of Jesus, God made physical, is God's stunning affirmation of the material world. Both Jesus' bodily resurrection, and the promise that we too will have physical resurrection bodies (1 Corinthians 15), continue to show how positively God sees material things. It is therefore deeply sub-biblical to say that the Christian message is about spiritual rather than material things.

3. 'Why bother to care for the earth – isn't God going to destroy it anyway?'

There are two main answers to this question. First, whatever God has in mind for the future, the task of Christians now is to be obedient to God's command to care for the earth (Genesis 1:26–28; 2:15). In one sense, it's none of our business if God wants to destroy what he's made – our job is to look after it until then! However, secondly, the idea that God might totally destroy the earth is actually built on very shaky biblical foundations. The fact is, whenever the Bible talks about the future of the earth it holds in tension the twin themes of destruction (judgment) and renewal (salvation). Often Christians have grabbed at one of these (usually destruction) and constructed a theology around it while totally ignoring passages that point in the other direction. A truly biblical understanding avoids both false extremes: the notion that God is going to destroy the earth completely, and the equally wrong idea that everything will slowly improve and evolve towards perfection. Rather, a balanced biblical understanding recognizes that God's judgment of all that is fallen, evil and sinful will mean a radical cleansing of the whole creation, but that God's saving love towards all he has made will eventually lead to the remaking, reshaping and renewal of creation.

4. Doesn't science contradict the Bible, especially when it comes to the creation accounts in Genesis 1 and 2?

True science can only ever inform and confirm God's word in the Bible, because good science is humanity seeking to explore and understand God's world. In fact Genesis 2, when God tells Adam to name each kind of animal, is the beginning of the science of taxonomy: identifying, differentiating and classifying are building-blocks of biology! Jesus' commands in Matthew 6 to study the birds and flowers are also an encouragement to us to see science as a way of 'thinking God's thoughts after him'.

However, there are difficulties when either scientists or Bible students try and push their discipline beyond its limits. Science is great at looking at how things work and change, but cannot answer the deeper questions of why we are here, why things work as they do, and why the universe is constructed in such a delicate, finely balanced way and seems able to keep itself in balance. The Bible, on the other hand, is often God's way of answering these 'why?' questions, but neither the whole Bible nor Genesis 1 – 2 is intended to be a straightforward scientific handbook on exactly how things have happened. Christians will always be split over whether God created the world in six twenty-four-hour days, or whether he used evolutionary processes over billions of years, but neither view must be an excuse for avoiding the main challenge of Genesis 1 – 2, to understand our dual nature as human beings: both made from 'the dust of the earth' as part of creation, and also called apart to be God's image in caring for creation.

5. God told us to 'fill the earth and subdue it' and to 'have dominion' over it. So, aren't the earth and its creatures simply there for our use and enjoyment?

This misunderstanding has often been present in Western Christian thinking, and has caused untold damage both to the planet and the reputation of the gospel. In fact, the Bible is very clear that it is God's world, not ours (Psalms 24:1; 50:10–11), and that it was created ultimately for Jesus (Colossians 1:16). We are permitted to use and enjoy creation as its tenants (Leviticus 25:23) and caretakers (Genesis 2:15), but not in a way that is careless, greedy or destructive. We are answerable to the owner: God. In fact the word in Genesis 1 for 'subdue' should be seen as meaning 'manage' or 'bring order to', and the word 'dominion' is about ruling over in a way that reflects God's gentle and just rule. In the light of Jesus, who came not to be served but to serve, we can describe this as servant kingship.

6. Shouldn't we be helping the poor, rather than worrying about wildlife?

It's a false distinction to separate caring for the poor from caring for the planet. God has made a world that is interdependent, where we as humans cannot survive without healthy ecosystems to give us food, water, shelter, clothing, fuel, and even the air we breathe. It is the world's poor who are suffering most from climate change and who are most directly dependent on the natural systems around them. Dr Stella Simiyu, a Kenyan botanist and member of A Rocha International's Council of Reference, puts it like this: 'The rural poor depend directly on the natural resource base. This is where their pharmacy is, this is where their supermarket is, this is in fact their fuel station, their power company, their water company. What would happen to you if these things were removed from your local neighbourhood? Therefore we really cannot afford not to invest in environmental conservation.'[2] It is also very important to remember that we should care for everything that God cares for – which includes the wildlife he made, sustains and entrusts to our care.

7. Doesn't the Bible tell us not to worry about tomorrow? Surely it is
God's job, not ours, to care for the planet?

Not worrying about tomorrow (Matthew 6:34) does not mean
not caring about tomorrow! Biblical faith is about depending on
God 100% to meet all our needs, yet at the same time taking up God's
call 100% to be co-workers in his kingdom. Somebody once put it: 'It's
my business to do God's business and it's his business to take care
of my business.' So, while God is committed to sustaining and caring
for the whole creation, he's delegated much of that to us! The story of
Noah is a great example: God didn't reach out and rescue all the
animals, he asked a human being to act on his behalf. He still does.

8. The problems are too big – what difference can I make?

It's easy to feel overwhelmed by the scale of the environmental crisis,
but here are a few thoughts that may help:

- Be local! Your responsibility is not to change the world on your
  own, but to 'become the change you want to see in the world'.[3]
  In other words, obedience to God's call rather than 'success' is
  what we're called to. Let's make sure we're changing what we
  can, and let God worry about the big picture.
- Get some perspective! 'Climate change is not one big,
  intractable problem but billions of tiny, tractable ones.'[4] In
  other words, if we break things down into everyday decisions
  we all make, together we can make a huge difference. As the
  riddle goes, 'How do you eat an elephant?' 'One bite at a time!'
- Be encouraged! World-changing movements can have small
  and seemingly insignificant beginnings. Think of William
  Wilberforce and the abolition of slavery, of Gandhi and the
  Quit India movement, or of how one travelling preacher in the
  Middle East 2,000 years ago, who died a 'failure', transformed
  the world.

9. Why a Christian environmental organization? Shouldn't we just
join with others?

Christians should indeed be involved in the wide range of conservation and environmental movements that already exist. However,
there is a key place for organizations such as A Rocha in educating

and challenging churches, and in linking the environment to clearly defined moral and spiritual values. Many conservation organizations have no worked-out idea of why obscure species matter. In addition, if we believe that caring for creation is part of seeking God's kingdom 'on earth as in heaven', then it should be as natural to have Christian environmental organizations as to have Christian relief and development agencies – they are both an expression of God's love through his people.

10. Wouldn't the planet be better off without people?

Many Christians might see this as a shocking question, yet it is being asked with increasing frequency. After all (it is argued), if we are responsible for all the problems, maybe the earth would look after itself better without us. Aren't we just the 'virus species'?[5] The evidence of the negative impact humanity is having today is clear, so Christians should be careful about simply affirming that we're in God's image and therefore the planet must be better off with us than without us. Rather we need to repent of failing to reflect God's image in how we treat the earth, and demonstrate by our actions that we can make a positive difference. If we believe God has entrusted creation to our care we need to be a lot more careful with it.

11. Haven't Christians got an appalling track record in caring for creation, and isn't that all based on what the Bible teaches?

There's no denying that Christians have often been guilty of allowing and even encouraging the misuse of God's creation. Many environmentalists see Genesis 1:26–28 ('subdue', 'have dominion', 'image of God') as putting humanity on a pedestal above other species and laying the foundation on which aggressive industrialization and unsustainable living have been built. However, there are three important things this ignores.

- It is not only Christianity that has been guilty of causing environmental disaster. Atheistic communism, aggressive secular capitalism and Islamic imperialism have all done the same. Any world-view that exalts human beings above other species, and forgets our interdependence with them, will lead to environmental disaster.

- *The track record of Christianity is far more positive than some have realized.* Alongside the many failures are inspiring examples of sustainable living and creation care, such as St Francis of Assisi, early Celtic Christianity, Benedictine monasticism in the Middle Ages and the Amish in America. It is when Christians have become captives of a human-centred culture, rather than allowing God's word to transform their culture, that greed, exploitation and carelessness have caused damage to creation.
- *The Bible does not teach that the world exists simply for humanity to use or abuse.* Every major theme of Scripture (and every chapter of this book!) shows us that God's world is precious and worthy of our care and respect. God himself is committed to sustaining and renewing creation, and has entrusted its care to humanity. Understood in context, Genesis 1 and 2 clearly teach that it is God's world (not ours!), that humanity is as much part of creation as called apart to be God's image, and that 'dominion' and 'ruling over' are about our exercising God's just and gentle rule: working to serve and preserve the earth and its creatures.

Notes

Introduction
1. N. T. Wright, 'Jerusalem in the New Testament', in P. W. L. Walker (ed.), *Jerusalem Past and Present in the Purposes of God* (Paternoster/Baker, 2nd edn, 1994), p. 70.

Chapter 1
1. Irenaeus, *Against Heresies* 4.20.1 and 5.6.1, in *Ante-Nicene Fathers, 1: The Apostolic Fathers with Justin Martyr and Irenaeus*, ed. Philip Schaff, available online at <http://www.ccel.org>.
2. Here are a few more examples; you might like to see how many more you can find. Psalms 19:1–6; 29:3–10; 33:5–11; 65:5–13; 74:12–17; 84:3–6; 89:9–13; 96:10–13; 97:1–6; 104; 135:6–7; 145:13–21; 147:4, 8–9, 15–18; 148.
3. Quoted in Martin Thompson, 'Living Colour', *CAM – Cambridge Alumni Magazine*, 49 (Michaelmas 2006), p. 18.
4. Read the stories in the book of Jonah and in Numbers 22:21–41.
5. Conservative Party Conference, Brighton, 14 October 1988, <http://www.margaretthatcher.org/speeches/displaydocument.asp?docid=107352>.
6. Lynn White, 'The historical roots of our ecologic crisis', *Science*, 155 (1967), pp. 1203–1207.

Chapter 2
1. This illustration is developed from a suggestion by C. J. H. Wright, *Living as the People of God* (IVP, 1984).
2. James Jones, Bishop of Liverpool, BBC Radio 4, *Sunday Morning Service*, 15 April 2001.

3. Press Release for A Rocha: Christians in Conservation (<http://www.arocha.org>) of which Sir Ghillean is a trustee, 30 March 2005.
4. M. Shellenberger and T. Nordhaus, *The Death of Environmentalism: Global Warming Politics in a Post-Environmental World* (Houghton Mifflin Co., 2007), p. 34. Available as a download at <http://www.thebreakthrough.org>.

Chapter 3
1. Russ Parker, *Healing Wounded History* (Darton, Longman & Todd, 2001), p. 8.
2. Bob Beckett with Rebecca Wagner Sytsema, *Commitment to Conquer* (Chosen Books, 1997), p. 53; Alistair Petrie, *Releasing Heaven on Earth: God's Principles for Restoring the Land* (Chosen Books, 2000), p. 31.
3. Walter Brueggemann, *The Land* (Fortress Press, 1977), p. 3.
4. C. J. H. Wright, *Living as the People of God* (IVP, 1983), pp. 37–38.
5. Parker, *Healing Wounded History*, p. 9.
6. Wright, *Living as the People of God*, p. 59.
7. *Living as the People of God*, p. 48.
8. Petrie, *Releasing Heaven on Earth*, pp. 203, 221. Almolonga is also featured in the first Transformations video, produced by the Sentinel Group; <http://www.sentinelgroup.org>.

Chapter 5
1. A. M. Allchin, *Bardsey: A Place of Pilgrimage* (privately published, 2nd edn, 2002).
2. Hans Küng, *On Being a Christian* (William Collins, 1977), p. 231.
3. Quoted in Wim Rietkerk, *The Future Great Planet Earth* (Good Books, 1989), p. 33.
4. In the New Testament, the Greek *stoicheia* usually means the basic principles that had rebelled against God (see Galatians 4:3, 8; Colossians 2:9, 20).
5. Ezekiel 47; John 7:37–38.
6. N. T. Wright, *New Heavens, New Earth* (Grove Booklets, 1999), p. 5.

Chapter 6
1. *Order for the Celebration of Holy Communion* (Church House Publishing), p. 165.
2. James Jones, *Jesus and the Earth* (SPCK, 2003), p. 12.
3. See chapter nine.

Chapter 7
1. Jürgen Moltmann, *God in Creation: An Ecological Doctrine of Creation*, The Gifford Lectures 1984–1985 (SCM Press, 1985), p. 71.
2. Zygmunt Bauman, *Globalization: The Human Consequences* (Columbia University Press, 1998).
3. *Sharing God's Planet* (Church House Publishing, 2005).

Chapter 8

1. Quoted in David Watson, *I Believe in the Church* (Hodder & Stoughton, 1985), p. 305.
2. See chapter nine for more information.
3. <http://www.livinglightly24-1.org.uk>.
4. <http://www.wen.org.uk/nappies/facts.htm>.
5. For instance, Exodus 20:8–11; Exodus 23:5, 12; Deuteronomy 25:4; Jonah 4:11.
6. <http://www.rainforestconcern.org/rainforest_facts/why_being_destroyed>.
7. Figures from the Aviation Environment Federation on an average annual mileage of 9,000 miles: <http://www.aef.org.uk/downloads/Howdoesairtravelcompare.doc>.
8. A Rocha has set up Climate Stewards (<http://www.climatestewards.net>) as an agency to offset polluting travel, aiming not only to soak up CO_2 emissions, but to improve biodiversity and work with local communities in developing countries.
9. <http://actonco2.direct.gov.uk>.

Chapter 9

1. Lausanne Covenant, <http://www.lausanne.org/lausanne-1974/lausanne-covenant.html>.
2. Archbishops of the Anglican Communion, *Five Marks of Mission* (1984, 1990).
3. See <http://www.arocha.org> for more details.
4. Interview with Rob Frost on *Care for Creation: The Biblical Basis*, A Rocha DVD.
5. Bishop James Jones, at John Ray Initiative Conference, Gloucester, 8 February 2003, at <http://jri.org.uk/resource/jesusearth_bishopliverpool.htm> (accessed October 2007).

Planet Whys

1. From a video interview with A Rocha conducted in 2005, available from A Rocha UK on a CD-ROM/DVD Environment Resource Pack.
2. Video interview, A Rocha, 2005.
3. Mahatma Gandhi, quoted on many websites, including <http://www.thinkexist.com/quotation> (accessed October 2007).
4. Nick Spencer and Robert White, *Christianity, Climate Change and Sustainable Living* (SPCK, 2007), p. 62.
5. 'Our viral like behaviour can be terminal both to the present biosphere and ourselves.' Paul Watson, in an essay 'The Beginning of the End for Life as We Know it on Planet Earth?' at <http://www.seashepherd.org> (accessed October 2007).

Where to go from here?
Resources to help you in caring for God's world

Planetwise Resource Pack: You've read the book, now get the pack! A CD-ROM and DVD packed with ideas to transform your worship, discipleship, lifestyle and mission. The pack has several short films (about A Rocha and about caring for creation), Bible study resources, sermon and service material, and practical ideas. Available for £10 including postage (made payable to 'A Rocha UK') from: A Rocha UK, 13 Avenue Road, Southall UB1 3BL.

Useful websites and organizations
A Rocha: <http://www.arocha.org>. The international conservation organization, working to care for God's world – and helping you to get involved.
Eco-Congregation: <http://www.ecocongregation.org>. Helping churches become greener. In England and Wales Eco-Congregation is managed by A Rocha UK.
John Ray Initiative: <http://www.jri.org.uk>. Connecting environment, science and Christianity.
Living Lightly 24:1: <http://www.livinglightly24-1.org.uk>. A Rocha's lifestyle challenge and website resource. There are links to dozens of other websites with practical environmental advice.

Operation Noah: <http://www.operationnoah.org>. Climate change campaign founded by Christian Ecology Link with support from British churches.

Books

This list does not aim to be in any way comprehensive, but is a guide to some books to take you further. The first three have one or more chapters by Dave Bookless in them.

Caring for Creation: Biblical and Theological Perspectives, edited by Sarah Tillett (Bible Reading Fellowship, 2005). A range of Christian scientists, biblical scholars and practitioners reflect biblically and share stories from A Rocha's work around the world.

When Enough is Enough: A Christian Framework for Environmental Sustainability, ed. R. J. Berry (Apollos, 2007). Overview of the whole subject theologically, scientifically, economically and more – a heavier read but worth the work!

Celebrating Community: God's Gift for Today's World, ed. Chris Edmondson and Emma Ineson (Darton, Longman and Todd, 2006). A series of reflections on God's call to live as community from people who've been involved with Lee Abbey – the Christian community which runs centres for the renewal of the church.

For the Beauty of the Earth, by Steven Bouma Prediger (Baker Academic, 2001). Well worth searching out. Aimed at Bible college students. A thorough and fully referenced look through the biblical call to care for creation and engage with the environmental debate.

Christianity, Climate Change and Sustainable Living, by Nick Spencer and Robert White (SPCK, 2007). An excellent summary of climate-change science accompanied by reflections on a biblical vision of 'the good life' and how that should affect our lifestyles now.

Glory Days: Living the Whole of Your Life for Jesus, by Julian Hardyman (IVP, 2006). The subtitle says it all.

The Busy Christian's Guide to Busyness, by Tim Chester (IVP, 2006). Not specifically about the environment, but a wise and helpful guide to re-ordering your priorities and lifestyle.

Kingfisher's Fire, by Peter Harris (Monarch, 2008). The thrilling and
challenging story of how A Rocha has grown from one project
to an international movement.

L is for Lifestyle, by Ruth Valerio (IVP, 2004). A practical and
challenging introduction for Christians to ethical living.

The Rough Guide to Ethical Living, by Duncan Clark (Rough Guides,
2006). Superb, easy to read and reliable guide to living in a way
that helps the poor and the planet. Not written from a Christian
viewpoint but most of it might have been!

Related titles from ivp

L
is for
lifestyle

Christian living that
doesn't cost the earth

You wake up and have a shower. The water is hot and the house warm. You pull on your designer-label trousers. You throw some bread in the toaster, watch the news on TV and have breakfast: coffee made with water boiled in the kettle and cereal with milk kept cool in the fridge. You scribble a note on a pad of yellow sticky notes and throw away the foil trays from last night's takeaway, before jumping in the car and setting off for work.

You have done nothing out of the ordinary, but already your lifestyle choices have had an impact on people and the environment right across the world.

So how can we live more responsibly? In this A–Z of lifestyle issues Ruth Valerio highlights the main threats to people and our planet, God's beloved creation. She shows how, by making small changes to our everyday behaviour, we can learn the secret of a life that is both fair and simple.

'Full of practical suggestions on how we can live in a way that is more consistent with the will of God our Saviour.'
Dewi Hughes
Theological Advisor, Tearfund

ISBN:
978-1-84474-025-3

Available from your local Christian bookshop or via our website at **www.ivpbooks.com**